CONSTRUCTION OF THE BUILT ENVIRONMENT

CREATIVE DESTRUCTION AND CONSTRUCTING THE BUILT ENVIRONMENT

From the first industrial revolution to the fourth

Gerard de Valence

First published in 2022 by Construction Economics Research, Kurmond, Australia.

constructioneconomicsresearch.com

ISBN 9780-0-6454977-1-7

Cover photo: Getty Images.

This book is typeset by Coverkitchen.
coverkitchen.com

Figures made by Rawtech Digital.
rawtechdigital.com

Construction
Economics Research

Table of Contents

Figures

Tables

Introduction

... craftsmen, mechanics, inventors, engineers, designers and scientists using tools, machines and knowledge to create and control a human-built world
Thomas Hughes[1]

I WAS ONCE ATTACKED by a colleague for, as he put it, 'not considering the great mass of people employed in construction'. We were working for a government inquiry into collusive tendering and discussing recommendations to improve productivity and efficiency in the final report. At the time there were significant changes affecting the Australian industry that had far more impact than the legislative and regulatory reforms the inquiry led to. The industrial relations system was moving from a centralised award based one to a more decentralised system with enterprise bargaining and site agreements. International contractors were entering the market and the larger engineering and architecture practices consolidating. As the industry began to recover from a speculative office building bubble and the economy rebounded from a deep recession, construction employment increased and continued to grow for the next few decades. Construction as used here refers to all the firms and organizations involved in design, construction, repair and maintenance of the built environment.

Where these longer run trends were going was not obvious at the time. There have been significant changes in the range of activities and types of firms involved in construction of the built environment over the last few decades. Two trends underpinning those changes were the increasing use of multi-disciplinary project teams as the boundaries between professional disciplines became less distinct,[2] and the inhouse versus outsourced decision about provision more common.[3] Facilities management is an example, an activity that used to be done internally but is now often outsourced, sometimes but not always to construction contractors. Consultants bid for work as contractors, and contractors do consultancy and project management. Urban planning was once primarily associated with design, but is now linked to real estate and development. The process of structural change in industry occurs as technology, institutional and firm capabilities develop and change over decades.[4]

One motivation for this book is a belief that the development of modern construction can provide a framework for understanding how current technological changes might impact firms and industries today. When considering the relationship between construction of the built environment and technological change the past is really the only guide available, so the starting point for this discussion is the first industrial revolution in England at the beginning of the nineteenth century when modern construction and its distinctive culture began to form, followed by the twentieth century's attempts to industrialise construction. This history is important because, after more than 200 years of development, construction of the built environment happens today within an established system of production based on a complex framework of rules, regulations, institutions, traditions and habits that have evolved over this long period of time.

But how useful is history and how can it be used? Are there appropriate historical examples or cases to study to see if there are lessons relevant to the present? The answers depend to a large extent on context, because a key characteristic of the history of technology is the importance of institutions and the political and social context of economic outcomes. [5] Also, understanding how policies were developed in the past and how effective they were requires understanding the changing context of policy implementation. However, as economist Paul Samuelson pointed out 'history doesn't tell its own story and 'conjectures based on theory and testing against data' are needed to uncover it.[6] Drawing the right lessons from history is a nuanced exercise.

Over time industries and products evolve and develop as their underlying knowledge base and technological capabilities increase. The starting point for a cycle of development is typically a major new invention, something that is significant enough to lead to fundamental changes in demand (the function, type and number of buildings), design (the opportunities new materials offer), or delivery (through project management). Major inventions give a 'technological shock' to an existing system of production, which leads to a transition period where incumbent firms have to adjust to the new business environment and new entrants appear to take advantage of the new technology.[7] Economist Joseph Schumpeter called this process creative destruction,[8] and it leads to the restructuring and eventually consolidation of industries. That is what happened to construction and related suppliers of professional services, materials and components after the first industrial revolution.

The drivers of development for industries in the twenty-first century are emerging technologies such as augmented reality, nanotechnology, machine intelligence, digital fabrication, robotics, automation, exoskeletons and possibly human augmentation. Collectively, these digital technologies are described as a fourth industrial revolution, and their capabilities can be expected to significantly improve as new applications and programs emerge with the development of intelligent machines trained in specific tasks.[9] Innovation and technological change is pushing against what are now long-established customs and practices of the industries in the diverse value chain that designs and delivers the projects that become the built environment.

How technological change affects these industries differs from more widely studied industries like computers, automobiles or aerospace because of the number and diversity of firms involved in designing, constructing and managing the built environment. With the range of separate industries these firms come from, construction of the built environment is the output of a broad industrial sector made up of over a dozen individual industries. Not an 'industry' narrowly defined, but a broad industrial sector that is organised into a system of production with distinctive characteristics.[10] A second difference is the age of these industries, many of which are mature industries in late stages of their life cycle.[11] These differences create a different context for questions about industry, innovation and technological change, about how firms compete and how the system of production is organised as fourth industrial revolution technologies like digital twins and drones spread through construction and the pace of digitization increases.[12]

Figure 1. Organization of the system of production

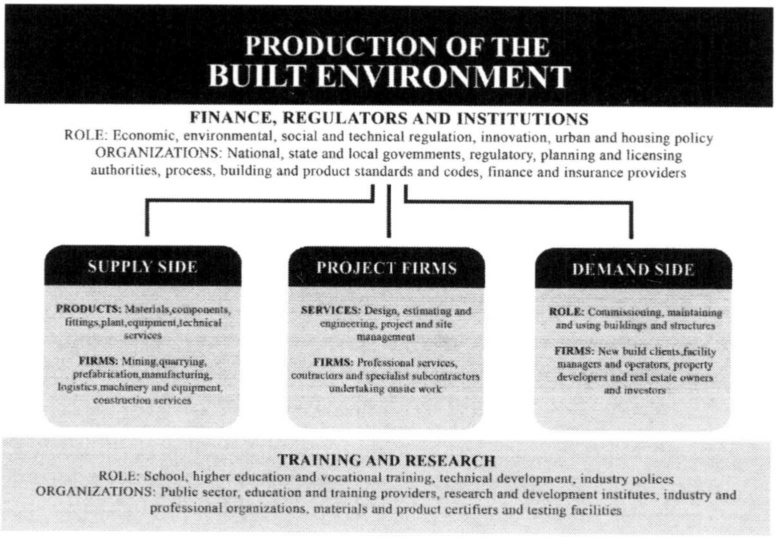

Source: Based on Gann, D. M. 2001. Putting academic ideas into practice: technological progress and the absorptive capacity of construction organizations, *Construction Management & Economics*, 19 (3), 321-30.[13]

As well as the contractors, subcontractors and suppliers for new builds, there are also many firms and people mainly engaged in the alteration, repair and maintenance of the built environment. The broad base of small firms is a distinctive feature of construction, and these family-owned firms engaged in repair and maintenance work will largely continue to use the materials and processes they are familiar with. Old technologies can survive long after the innovations that eventually replace them arrived, such as the telegraph, fax machine and vinyl records with telephones, email and CDs.[14] Stone, tile, brick and wood have been widely used materials for millennia, and industrialized materials like corrugated iron and concrete are ubiquitous. For maintaining and repairing the existing stock of buildings and structures, many of the skills, technologies and materials found today will continue to be used far into the future. That does not mean firms mainly involved in repair and maintenance will not be affected in some way by the fourth industrial revolution.[15]

Figure 2. Functions of the built environment

FUNCTION	PAST	PRESENT	FUTURE
Accommodation : Where we stay	One of the characteristics of construction is that a large part of the present industry maintains the legacy of the past industry		
Workplace : Where we make things			
Services : Where we get things			
Logistics : How things get to us			
Infrastructure : How it works			

Construction of the built environment has characteristic organizational and institutional features because it is project-based with complex professional and contractual relationships.[16] How firms utilise technology and develop technological capabilities differentiates them within this location-based system of production. Emerging technologies in design, fabrication and control have the potential to transform construction over the next few decades, possibly less, and the book suggests firms will follow low, medium or high-tech technological trajectories, determined by their investment in the emerging technologies of the fourth industrial revolution.[17]

There are, however, few specific predictions, beyond a broad view of what future construction might look like. That view is based on successful solutions being found for the many institutional and technical problems involved in transferring fourth industrial revolution technologies to construction. Without downplaying the difficulty of those problems, similar challenges have been met in the past, but those solutions led in turn to a reorganization of the system of production.

There are very many possible futures that could unfold over the next few decades as technologies like artificial intelligence (AI), automation and robotics develop.[18] However, the key technology underpinning these further developments is intelligent machines operating in a connected but parallel digital world with varying degrees of autonomy. These are machines that have been trained to use data in specific but limited ways, turning data into information to interact with each other and work with humans. The tools, techniques and data sets needed for machine learning are becoming more accessible for experiment and model building,[19] and new products like generative design for buildings plans, drone monitoring of onsite work and 3D concrete printers are available.

Intelligent machines are moving from controlled environments, like car manufacturing or social media, to unpredictable environments, like driving a truck. In many cases, like remote trucks and trains on mining sites, the operations are run as a partnership between humans and machines. There are also autonomous machines like autopilots in aircraft and the Mars rovers. As well as rapid development of machine intelligence, technological change in the form of new materials, new production processes and organizational systems is also happening. Sensors and scanners are widely used, 3D concrete printing is no longer experimental, cloud-based digital twins are available as a service, and online platforms coordinate design, manufacture and delivery of building components using digital twins.

The book argues a period of restructuring of construction similar to the one that occurred in the second half of the 1800s is about to start. That was when the new industrial materials of glass, steel and reinforced concrete arrived, bringing with them new business models, new entrants and an expanded range of possibilities.[20] The development of modern construction was not, however, a smooth upward path of progress and betterment. It went in fits and starts as new inventions and innovations arrived, slowly then quickly, often against critics of the modern system of production and workers, fearing technological unemployment and lack of government support during a time of technological transition, who resisted new technology and sometimes sabotaged equipment.[21] The issue in the past, like today, was in fact not the availability of jobs but the quality of skills during the diffusion of new technologies through industry.

The only previous comparable period of disruptive technological change in construction of the built environment is the second half of the nineteenth century. Between 1850 and 1900 construction saw the rise of large, international contractors, who reorganized project management and delivery around steam powered machinery and equipment. In particular, the disruptive new technologies of steel, glass and concrete, which came together in the last decades of the century, led to fundamental changes in both processes and products. If that is any guide, we can expect technological changes to operate today over the same three areas of industrialization of production, mechanization of work, and organization of projects that they did then. And today, just as in 1820 when no-one knew how different construction would be and what industry would look like in 1900, we can't see construction in 2100. That is a long way out, and we can only guess at the level of future technology. We can, however, use what we already know from both history and the present to form a view of what is possible over the next few decades based on what is currently understood to be technologically feasible.

Decarbonisation will be another challenge in the near future for construction of the built environment. Here industry responses to changes in consumer demand is significant, for household heat pumps and rooftop solar for example, but governments have the important role through their building standards and codes, such as solar mandates for new buildings and building codes that require full electrification of new housing. Adapting to climate change by retrofitting the built environment and making cities and infrastructure more resilient will be done by construction firms within a detailed and complex regulatory and policy environment and requires targeted industry policies to be successful.

It should be clear that the role of the technologies discussed here will be to augment human labour in construction of the built environment, not replace it. Generative design software does not replace architects or engineers. Optimization of logistics or maintenance by AI does not replace mechanics. Onsite construction is a project-based activity using standardized components to deliver a specific building or structure in a specific location. The nature of a construction site means automated machinery and equipment will have to be constantly monitored and managed by people, with many of their current skills still relevant but applied in a different way. Nevertheless, in the various forms that building information models, digital twins, AI, 3D printing, digital fabrication and procurement platforms take on their way to the construction site, they will become central to many of the tasks and activities involved. Education and training pathways and industry policies with incentives for labour-friendly technology will be needed.[22]

Because construction involves so many firms and people the technology driven changes discussed here will have significant and profound economic and social consequences. This would be a good opportunity for government and industry to work together to develop policies and roadmaps for those firms, and to support 'the great mass of people' employed in construction of the built environment who will be affected by them. The future is not determined, although technological change and creative destruction continue to reshape and restructure industry and the economy, decisions made today create the future.

Outline of the Book

Chapter one introduces two ideas that provide the framework for the book: general purpose technologies (GPTs) and Joseph Schumpeter's idea of creative destruction.[23] Major new technologies are not common.[24] Although GPTs are rare they are powerful. As well as becoming widely used by existing industries they create new industries around new products and services. Schumpeter described the ongoing process of structural change in the economy, as the contributions of different industries to total output rises or falls over time, as creative destruction. Through diffusion of new technologies new industries are created and established industries face new entrants and have to restructure. For Schumpeter this was technological progress, driven by new goods and new production processes, with competition between firms not about costs and prices but research and development (R&D) that leads to new products, new methods of production and new forms of organization. Successful firms grow and thrive, unsuccessful firms are taken over or fail.

The following chapter looks at the origins of the modern system of construction of the built environment. Chapter two is on the emergence of general contractors and professional services during the transition from the craft system of production to an industrialized system based on competitive tendering with documentation for design, estimates and forms of contract. At the same time as these changes in the method of procurement were happening, the scale and scope of construction projects was increasing, new types of buildings were required by clients, and work was being mechanised. This history shows how the modern industry developed in the United Kingdom (UK), responding to changes on both the demand side, for new types of buildings, and on the supply side, with new materials and methods.

Why would be experience of the industry so many years ago be relevant today? There are two parts to the answer. The first is that the late nineteenth century is the only other period of comparable disruptive technological change we have. The second is that the effects of technological change on industry structure and performance might again be in the same key areas as industrialization, mechanization and organization of projects and processes, but in the twenty-first century these effects will be heightened and quickened by the network effects associated with digital platforms and AI. Because industry structure (the number and size of firms) is fundamentally determined by technology,[25] the emergence of new technologies and periods of rapid change can lead to new industries. Such creative destruction will also extensively restructure existing industries.

Chapter three starts with the three dimensions of technological change in nineteenth century construction identified by engineering historian Thomas Peters: construction was industrialized with standard components and mass production using new materials like steel, plate glass and plastics; sites were mechanized with steam powered railways, cranes and excavators; and new forms of project management were required to maximise efficiency of the machinery.[26] Fourth industrial revolution technologies can be expected to work along these same dimensions in the twenty-first century.

The chapter then introduces Thomas Hughes' industry life cycle, which he based on his study of the development of electricity generation, as a method of analysing the processes underlying the role of invention, innovation and new technology in the evolution of an industry over time.[27] The stages and cycles of the life-cycle are applied to construction and the role of invention and innovation in those stages discussed, as are the characteristics of a long-established industrial sector identified by Hughes. Examples of the effects of GPTs and the role of incremental innovation are given.

Because construction requires inputs from materials, manufacturing and professional service firms, these industries have been included in the discussion about how technological change has affected the industries involved in production of the built environment since the first industrial revolution. More than two hundred years later construction of the built environment is a technocratic system of production, based around standards and codes, contractually defined professional roles, and with a high degree of technological lock in due to the culture, age and complexity of the system. The 'embeddedness' of the construction technological system is found inside the professional institutes and organizations, trade and industry associations, government regulations and licensing, standards and codes, insurance and finance providers and regulators.

Chapter four is on modern construction and industrialized building, or offsite production of components, pods or modules. The technological base of offsite manufacturing is a mix of those from the first, second and third industrial revolutions, like factories, computers and lean production. This has not become a widely used system of production – manufactured buildings have succeeded in specific but limited markets. Instead, onsite construction continues to have a deep, diverse and specialised value chain that resists integration because it is flexible and adapted to economic variability. The chapter reviews progress on offsite production in construction, followed by four short studies of industrialized building, The three cases of Sears Modern Homes, Legal & General manufactured houses and Katerra had different problems: two failed, and the 1990s Japanese automated building systems were only used for a small number of buildings. The chapter then looks at the development and potential of software platforms for integration of procurement, design, and manufacturing in modern methods of construction.

Chapter five begins with an outline of the range of technologies that are included in what has become known as Industry 4.0, and its close relative Construction 4.0. Progress is found at the technological frontier, where startup firms apply the tools and techniques of the fourth industrial revolution. Funding for construction tech startups began to increase in 2018 and examples of funding, firms and products are provided, outlining the technological frontier in construction of the built environment in the early 2020s. Two technologies are discussed in more detail.

The first is artificial intelligence. By combining, managing and integrating data from many sources with analytical and machine learning capabilities, AI can make reliable predictions about the state of the world.[28] Applications of this form of task oriented narrow AI in construction are generative design, the operation and maintenance of plant and equipment, daily progress on a construction site or monitoring the condition of a building. Although there are technical challenges involved in applying machine intelligence, it is not unrealistic to think they will be solved as the capabilities of deep learning and cloud-computing improve and increasing digitization provides more construction-related data.

The second is additive manufacturing (3D printing) and digital fabrication.[29] There are now dozens of 3D concrete printing machines available, ranging in size from desktop printers to gantry systems that can build three and four stories. Suppliers offering manufacturing on demand with print farms (factories with many machines) makes local production of many building components possible, with an onsite or nearsite fab producing many of the concrete, metal, plastic and ceramic fittings and fixtures for a building. This does not suggest the end of mass production of standardised components – economies of scale are the economic equivalent of gravity. However, the potential effects of onsite and nearsite production on the role of contractors and the organization of projects are significant. Underpinning digital construction are digital twins of projects using building information modelling (BIM). The chapter closes with an argument for government policies promoting digital construction with BIM mandates.

Industry policy is the subject of Chapter six.[30] Governments can have major impacts through regulation, tax, education, training, innovation and R&D policies, and through purchasing policies. Public policies specifically for construction of the built environment are also subject to the effects of policies for contentious issues like housing and infrastructure development. The two policy areas discussed are BIM mandates and building standards and codes. The experience of the UK and the US are used as examples, with a discussion of the construction reform movement's promotion of offsite manufacturing in the decades before the new UK industrial strategy with a BIM mandate was launched in 2011.

Another area where governments can promote industry development is through building standards and codes. Building characteristics like materials, access, ventilation and fire safety are regulated by standards and codes, and contractor accreditation for standards is often required by clients. The performance of the built environment is to a large degree measured against the baselines set by standards for health and safety, environmental management and process control. Although agreeing new standards is a lengthy process, they are universally accepted and applied because of the rigorous scientific and engineering research they are based on. Therefore, an important element in a strategy to increase innovation in construction of the built environment is to increase funding for testing laboratories.

Decarbonisation of both the construction and operation of the built environment will be crucial in reducing and eliminating greenhouse gas emissions. Building energy codes provide a tool for governments to mandate the construction and maintenance of low-energy buildings. To do this, the energy use of buildings must be monitored and managed, and buildings must be built and retrofitted to use less energy. A built environment carbon budget is required, and a global standard for determining greenhouse gas emissions for cities is under development. Examples of the effectiveness of energy codes are discussed.

In chapter seven, three technology adoption and implementation trajectories for construction over the next few decades are discussed. These trajectories can be low-tech, medium-tech and high-tech. What differentiates the three is firstly the investment by firms in development, and secondly which new technologies are taken up, which leads to the different trajectories for firms. On the low-tech path firms continue with business as usual. They are followers not technological leaders, and change is slow across the large number of these small local firms. This part of construction continues on a path of incremental innovation, similar to the present, becoming smarter and better at managing and using information as time goes on.

Firms on a medium tech path are upgraded and modified. These firms invest considerably more in technological development, making significant choices on which technologies to pursue and invest in. Those technologies, in turn, require changes to the way firms are organised and the way they organise their projects. Some businesses are much better at this than others. The companies on this path are technology leaders, using digital twins and developing platforms for their supply chains. The high-tech path outlined is a hybrid production system based around AI and digital fabrication about a decade in the future. New production technologies automate many tasks and materials, and machinery becomes smart and networked with embedded processors. Humans partner with machine intelligence to accomplish many tasks, and use robots or exoskeletons for most physical work, with remote control of automated or semi-autonomous plant and equipment, while fabricated and modular components combine with automated systems and specialised onsite assembly robots to transform the building process.

Despite the extent of technological change expected over the next few decades, some new industry will not appear to undertake construction of the built environment. However, that does not mean it will not undergo creative destruction as a result of future, but foreseeable, developments in AI, automation and robotics. Economies grow by upgrading the products and services they produce and export, but the technology, capital, institutions and skills needed to make new products are found in related products with common labour and capital requirements. This network of relatedness between products means industries develop goods and services close to those currently produced.[31] However, with the wide range of new production technologies currently emerging, such as 3D printing of concrete, automated machinery and materials like engineered wood, construction of the built environment is a laboratory for the fourth industrial revolution. Because it is not possible to know now which of these technologies will work at scale, the role of industry policy as facilitator is to promote digitization and decarbonization through revised building standards and codes, and to provide opportunities for new methods of production, organization and management to be tested and trialled on projects.

The Conclusion in chapter eight argues the technologies discussed like building information modelling (BIM), digital twins, AI, digital fabrication and procurement platforms will become central to many of the tasks and activities involved in construction of the built environment. While this might take a decade or more, as these fourth industrial revolution technologies become more competitive and their knowledge base deepens, the development path taken in construction will be distinct and different from the path taken in other industries. This path dependence varies not just from industry to industry, but from firm to firm as well.

There is a technical Appendix on measuring the economic role of built environment industries with two proposals for revising construction statistics. The data from national statistical agencies using the *System of Industrial Classification* can be used to measure output and employment of the industries involved in construction of the built environment. More accurate and complete data would be provided by a built environment sector satellite account, which could be regularly produced by national statistical agencies as for other sectors like tourism and agriculture. Finally, the current SIC structure of construction statistics should be revised to incorporate the work done by construction trades into the main categories of residential building, non-residential building and engineering construction.

1 Industrial Revolutions and Creative Destruction

Only when we observe over decades do we see the arrangements and processes that form the economy coming into being, interacting, and collapsing back again. Only in the longer reaches of time do we see this continual creation and re-creation of the economy.

Brian Arthur[32]

INDUSTRIES GO THROUGH periods of growth, competition and restructuring as the technology they are based on goes through cycles of invention and innovation, and this process of creative destruction separates firms that are successful adopters of new technology from firms that fail. As a result, technological change is not constant, but proceeds in fits and starts, with bursts of new products and processes when new technologies arrive, followed by periods of refinement and incremental improvement. This process is clearly seen in industries like electronics and automobiles, but it also affects industries involved in construction of the built environment.

Construction involves many firms and organizations in the design, construction and maintenance of the built environment, and this construction value chain and the way it is organised into a system of production is extraordinarily wide, deep and complex. Firms from many different industries participate, together making up a large industrial sector linking mining, manufacturing, professional services, contractors and subcontractors.[33] One reason why innovation is hard to see clearly in construction is because so much happens in these related industries rather than onsite. And today, across those built environment industries, different parts of the digital construction system are in different stages of development.

The arrival of new technologies that lead to periods of rapid change is not common.[34] They are rare but powerful, because they apply to and are used by many industries in many different ways. Steam power was such a general purpose technology (GPT), and is one of the three GPTs originally identified by economist Paul David in 1990 as the drivers of the first, second and third industrial revolutions: steam, electricity and information technology (IT).[35] It is widely believed AI is a new GPT,[36] and one that may be as disruptive as steam power was in the nineteenth century, when contractors replaced master builders. Such restructurings happened to construction three times over three centuries as new GPTs were adopted.

Figure 3. Four industrial revolutions since 1800

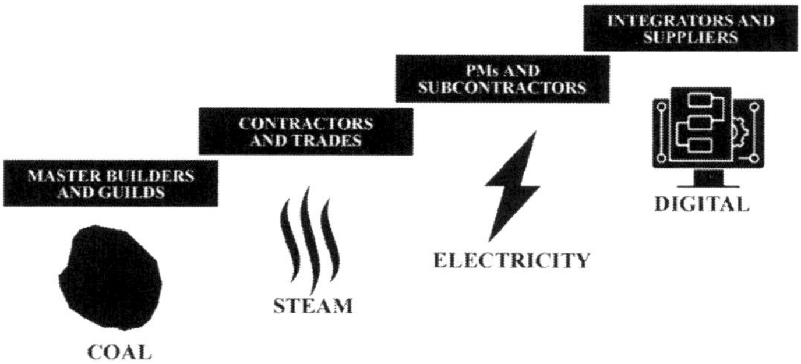

The three previous industrial revolutions were driven by new energy sources such as coal and steam in the nineteenth and electricity in the twentieth century. Over this long time, the structure of construction evolved through two stages, first from master builders and craft guilds to contractors and tradesmen, then to the modern project manager (PM) and subcontractor structure. Compared to steam power, the impacts of the other GPTs on construction of the built environment have been less dramatic as they did not significantly affect industry structure or dynamics. Electricity did not affect onsite construction in the same way it did manufacturing, which needed to reconfigure factory layouts, because onsite steam powered machines were generally replaced by petrol and diesel ones doing the same work. Computers and IT have restructured office work everywhere, and so far have affected industries like retailing, travel and entertainment far more than construction.[37]

GPTs like electricity, computers and the internal combustion engine became widely used across industry without significantly restructuring and reorganizing construction in the way steam powered mechanization did, because they essentially upgraded existing capabilities. Construction adopted these new GPTs and used them to improve efficiency, but they did not require a major change in the form of industry organization that emerged during industrialization and mechanization in the nineteenth century. Neither electricity nor computing had a significant effect on the organization of construction because the evolution from contractors and trades to PMs and subcontractors was not driven by those technologies. However, the change from master builders and crafts to contractors and trades was a break from the past, and the result of industrialization and mechanization.[38]

Concepts like an industrial revolution or GPT are too broad to have much explanatory power. At any one time there will be a number of GPTs at different stages of their life cycle, because the use of a GPT takes decades of development for its use to become widespread throughout the economy.[39] However, the key feature of a GPT is its 'pervasiveness' as it spreads and diffuses and becomes used by many other sectors in the economy, which leads to 'complementary investments and technical change' from the users.[40] The internal combustion engine and computers are good examples of GPTs with large industrial sectors and many specialised firms in a complex system of production. Historian Thomas Hughes called these industrial sectors 'technological systems', to describe industries that form a dense network of producers, suppliers and materials, saying 'Technological systems solve problems or fulfill goals using whatever means are available and appropriate; the problems have to do mostly with reordering the physical world in ways considered useful or desirable, at least by those designing or employing a technological system'.[41]

In *The Nature of Technology* economist Brian Arthur describes technology as combinations of assemblies or modules in a hierarchy of systems and sub-systems, each of which is a technology that solves a problem required for the whole system to work. He concludes there is no characteristic scale for technology, it works at all scales and this distinguishes it from products and processes. He then develops a theory on what (physical effects) and how (measurement and observation) is being combined and recombined as technology develops.[42] Because most new inventions are based on some new combination of existing technology, as more knowledge, new materials and equipment and so on accumulate, the range and number of possible new inventions increases exponentially. Technology development is a dynamic and fluid process of continual improvement as modules with new combinations of technology replace older, less functional, ones.[43]

Perennial Gales of Creative Destruction

Economist Joseph Schumpeter used 'perennial gales of creative destruction' to describe the long-term effects of GPTs and technological change. At any one time there are growing sunrise industries and declining sunset industries in the economy, an ongoing process of structural change as the contribution to total output of individual sectors rises or falls. The creative part is the invention and innovation that leads to new products and services, the destructive part is the demise of less successful firms and the decline of industries as they become less relevant.

The starting point for a cycle of creative destruction is a powerful new general GPT that is not only used by existing industries but creates new industries. These evolve and develop as the underlying knowledge base and technological capabilities increase and become more complex.[44] If, after a period of development, this GPT gives a technological shock to an existing system of production, it leads to a transition period where the firms involved have to adjust to a new business environment, which in turn leads to a restructuring and consolidation of the industry.[45] This is what happened to construction in the second half of the nineteenth century, with iron-framed and steel-reinforced concrete buildings industry had to not only master the use of these new materials, but also develop the processes and project management skills the new technology required.[46]

Schumpeter saw economic development as a dynamic process, occurring discontinuously over time, driven by waves of product innovation that sweep away old industries producing old goods and services. Firms search out new goods and new production processes to increase profits, so competition between firms is not only based on costs and prices but also the R&D that leads to new products, new methods of production and new forms of organization. For Schumpeter this was technological progress. Because of the search for profits, another innovation will come along, allowing the development process to restart as the cycle repeats itself.

The clustering of innovations and their cascading effect on economic development was also at the core of Schumpeter's business cycle theory.[47] At the time he was writing, those waves of innovation had been driven by the key technologies of steam power, the internal combustion engine and electricity.

When a phrase or term becomes part of popular language it is often used a shorthand way of referring to a particular set of ideas, for example the term 'big bang' policies, or attaching 'gate' to political scandals. The ideas may be only loosely connected to the phrase, and its application may or may not be appropriate, but a shared understanding of the basic concept and widespread use gives it a life of its own. 'Creative destruction' is one of those phrases, now generally understood to be a process where current firms and their products are replaced over time by new firms with new products. It has become a way to describe the effect of innovation and new technology on the economy. Historically, it was not uncommon for a technology at its most advanced level of development to be overwhelmed by a new, cheaper or better technology. For example, clipper ships and steam ships in the late nineteenth century, and diesel locomotives being replaced by electric ones in the mid-twentieth century.

Earlier economists like Adam Smith and Alfred Marshall had understood the driver of industrial change and long-run economic growth is technological development, but their models of the economy assumed all firms were small, had the same access to technology, and all markets were competitive. Schumpeter thought their models were incomplete because they did not include the processes of invention and innovation and their effect on firms and industrial structure. He argued successful firms would innovate more, grow faster, become larger and, if possible, use the new technology to acquire market power and dominate a market as a 'large scale establishment'. Schumpeter's model was based on two key propositions: there is a relationship between the size of a firm and innovation, with most innovation done by a relatively small number of large firms; and there is a relationship between innovative activity and a firm's market power, which is the extent of control a firm has over product and supplier prices.

Schumpeter argued the act of investment creates knowledge that can be used by both the original investor and other firms to make better informed decisions about future investments. The 'leading sectors' in this model are the ones which generate the most externalities and where 'spillovers' and diffusion of technologies to other industries can occur most rapidly. The investment linkages central to Schumpeter's model are also in 'learning-by-doing' models that stressed the importance of knowledge becoming a function of total capital stock. However, in those models technological change was a by-product of investment, rather than the outcome of deliberate activity by entrepreneurs. In Schumpeter's model of the economy the main agents for change are entrepreneurs, who see that improvements, change, and progress can be made, compared to capitalists who bear the financial risks, inventors who have the ideas, and managers who are concerned with production and distribution.

TECHNOLOGY, INNOVATION AND DIFFUSION

Two other elements of Schumpeter's model are important. The first is the role of economic incentives in the development of new technology, which requires investment in R&D, innovation and intellectual property. If firms cannot capture the benefits of innovation and intellectual property (IP) for some reason, because of imitation, piracy or secure supply of materials for example, they will not invest in innovation. Technology and research and development have significant costs associated with them, and Schumpeter recognised that investment in R&D was a calculated risk for a firm but, if successful, would benefit society as a whole. The model is fundamentally based on the rate of return in determining investment in innovation and the role of firms in deciding to invest in innovative activity. In his study of technology and industry economist Nathan Rosenberg described this relationship:

> Powerful economic impulses are shaping, directing and constraining the scientific enterprise. These impulses are rooted in two facts: First, scientific research is costly activity; second, it can be directed in ways that may yield large economic rewards. Industrial societies have created a vast technological realm that is very closely shaped by economic needs and incentives. This technological realm, in turn, provides numerous ways in which daily economic life has become closely linked with science. That realm defines the directions that promise large financial rewards and provides many problems and empirical observations that stimulate scientific research.[48]

The second element that is important is the time taken for diffusion of an innovation, as its use spreads among firms across the economy and industries. The introduction of an innovation by one firm leads to diffusion of the knowledge and makes it easier for other firms to make decisions about their own production, known as spillovers. How and why a new technology spreads through the economy and society is determined by many factors, however studies of historical cases such as tractors, electricity, TV and phones have given good examples of technology diffusion and the time taken because parallel changes in forms of organization, methods of production and patterns of consumption are required. These are not decisions firms and households make lightly, and it takes 15 to 30 years for a new technology to reach 90 percent of its potential market.[49]

The role of new technology in economic growth is crucial because new investment does not merely reduplicate old capital but embodies technical progress, so investment shifts production to a new level. The act of investment in innovation and human and physical capital is important for generating the externalities that lead to growth and further innovation. The introduction of an innovation by one firm leads to diffusion of the knowledge and affects other firm's decisions about their own production possibilities.[50] Eventually the pace of diffusion associated with that particular innovation will slacken and so will the rate of economic development in the process first described by Schumpeter. Because of the search for profits another innovation comes along, allowing the cycle to keep repeating itself.

A successful innovation will develop into an increasingly large market that will attract many new entrants, however, not all those firms will survive. In a typical industry life-cycle the growth phase is followed by a consolidation phase, where the largest firms take over or merge with others in the industry and unsuccessful firms exit. A mature, stable industry has few new entrants and few large established firms rarely fail. Automobiles provide a good example. There were hundreds of manufacturers in the early 1900s with high entry and high exit, reduced to three in the US by the 1930s. Today globally there are about a dozen large firms (VW, Toyota, Ford, GM etc.) that largely assemble components from a couple of dozen specialised manufacturers (engines, wiring, dashboards, seats etc.). With the innovation of electric vehicles a new cycle is beginning with new entrants (Tesla, BYD, Rivian etc.) and new products.

Figure 4. Innovation and the industry life-cycle

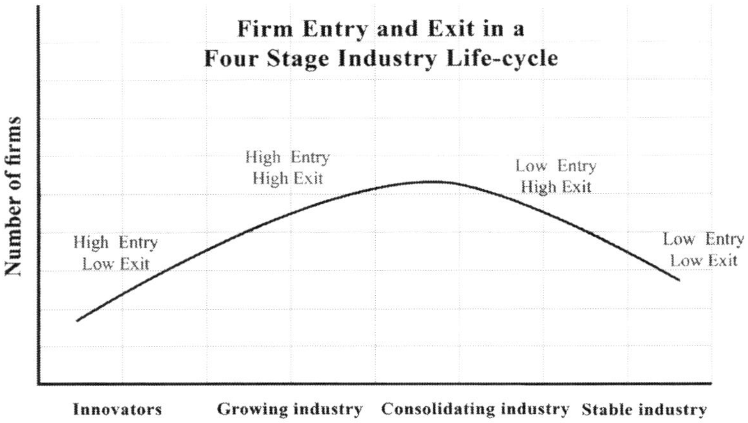

In one way or another, everything that has been written about innovation and technological change over the last 80 years addressed some aspect of Schumpeter's model of industry development. These decades saw rapid development of high-tech industries like aerospace, automobiles, pharmaceuticals, computing hardware and software, and telecommunications, so his influence on the way innovation and technological change is researched and analysed has been profound. In those industries the drivers, characteristics and dynamics of technological change have been the subject of intense interest and refinement.[51] Innovation is now understood to be radical, incremental or disruptive, technological change can be embodied in physical capital or disembodied as intellectual property. The diffusion of technologies like the internal combustion engine and computers takes at least a couple of decades and they are taken up at different rates in different industries, depending on where an industry is in the life-cycle.[52]

The development of high-tech industries in the second half of the twentieth century was extensively studied and documented, especially the computing hardware and software industries. These industries have a few large firms and a dominant technology that has gone through several generations over a few decades, making them ideal case studies of firm, product and process life cycles. Major manufacturing industries are similar, where detailed national data based on the *System of Industrial Classification* (SIC) allowed structural change in manufacturing to be researched, linking innovation to SIC codes and investment in R&D, patents and locations. Some industries, however, are more difficult subjects because they are not concentrated in a few firms or a few places, or do not fit neatly into the SIC codes. The contractors, subcontractors, manufacturers and professional services firms involved in construction of the built environment are a good example of a production system made up of industries where the data is spread across many different SIC classifications used by national statistical agencies. Innovation is harder to see and study in a system of production that is geographically dispersed and project based.[53]

Innovating in a complex, long established industrial sector like construction of the built environment can also be difficult. The institutional architecture can impose regulatory hurdles or other policy disadvantages on new technologies, and government expenditures often support existing technology. Lenders are risk averse. There are subsidies and price structures that favour incumbents and ignore externalities like the environment and public health. Educational curricula, career paths and professional standards are oriented to existing technology.[54] The dominance of existing technologies is further reinforced by imperfections in the market for technology such as network economies, lumpiness, split incentives and the need for collective action.[55] New technologies have to lower costs and improve performance to successfully compete with existing technologies and vested interests in the current system of production of the built environment.

Therefore, the effects of new technology on a large, long-established industrial sector takes time and will be uneven.[56] Nevertheless, AI, 3D printing and digital fabrication are two examples of major new GPTs that have profound economic implications for the organization of construction of the built environment. The impacts of new technology on a developed and embedded system of production like construction are often thought to be gradual, changing industry practices over time without significantly affecting their structure or dynamics.[57] There are good reasons to think this may be wrong because of advances in machine learning and automation, and the broadening potential of 3D printing and digital fabrication.

It is possible, or probable, that AI will lead to a reorganizing of construction from one centred on project managers to one based on AI powered platforms providing cheap, outsourced business processes and integrators that combine new methods of onsite and offsite production.[58] For both industry majors and smaller firms, a cycle of creative destruction appears to have begun, with rapidly changing technology offering many possibilities, opportunities and threats. This new cycle is starting more than 200 years after the emergence of modern construction in the first industrial revolution.

2 Continuity and Change in Construction after 1800

Bidding is as much a human game as it is a technical science
Howard Davis[59]

OVER THE COURSE of the nineteenth century traditional craft construction was reorganised and restructured around competitive bidding, general contractors and the new professions of architects, engineers and quantity surveyors. This replacement of one system of production by another one that is more advanced technologically is creative destruction at work. As described in the following chapters, this was a drawn out process driven on one hand by innovations in building procurement and project management, and on the other by the new industrialized building materials of iron, steel, glass and concrete.

There were fundamental changes in the procurement of construction during the first and second industrial revolutions, from the early system of craft production in the seventeenth and eighteenth centuries to the emergence of the general contractor and the professions with modern methods of procurement. Procurement here is the commissioning or purchasing of buildings and structures and any associated supplies and services required, by a client that pays for the work. A historical overview is useful because there is also a surprising, and not always recognized, continuity in the methods used and problems found.[60]

The process of designing and delivering buildings is as old as civilization. However, the procurement of projects has not attracted as much attention as the buildings and structures themselves, or the organization of the work, because in the distant past this was done by imperial edict for essential granaries, armouries, pyramids or temples. Procurement itself dates back to a red clay tablet found in Syria, dated between 2400 and 2800 BCE. This earliest procurement order was for '50 jars of fragrant smooth oil for 600 small weight in grain'.[61]

The earliest records on construction procurement are from Rome in the middle of the millennium around 500 BCE. What is significant about the history of procurement is the continuity of the basic characteristics of the methods used since then. Plutarch wrote on construction where 'artists' submit estimates and drawings and 'they select the one who, at the lowest price, promises the best and quickest execution'. It was common to divide large projects into work packages, with the Long Walls of Athens broken into 10 and the Roman Coliseum into four separate contracts.[62]

Along with continuity of procurement methods, issues around client and contractor performance are also familiar. The stereotypes of capricious and poorly informed clients and duplicitous contractors have a long history. For example, Marshall Vauban was a military engineer and builder of fortifications for the French monarchy who insisted suppliers should be selected on quality, not just on price. In a letter from 1685 to his Minister he complained of delays due to budget cuts and argued:

> Breaking of contracts, failures to honour verbal agreements and new adjudications, only serve to attract those firms which do not know which way to turn, rogues and ignoramuses, and to make those with the knowledge and capability of directing firms, beat hasty retreats. I would add that they delay and inflate considerably the cost of these works, which are the worst since these cuts and the cheapness sought are imaginary. For the contractor is ruined ... He does not pay the merchants who supply the materials, pays badly his employees, cheats on those he can, has only the worst, and since he is cheaper than the others, uses the poorest materials, quibbles about everything and is always crying for mercy ... go back to plain dealing; pay the price for the works and do not deny an honest salary to a constructor who fulfils his duties; that will always be the best deal you can find.[63]

This history of procurement describes the development of new methods for organising construction in England in the nineteenth century, where many of the major projects of the time were built. These developments are relevant to many countries today because of the influence England had on the forms of our modern language, laws, institutions, and governance.

While ideas like competitive tendering, enforceable contracts, subcontracting and measurement of costs with a bill of quantities are now widespread and common, this was not the case 200 years ago and these innovations came with the first industrial revolution that began in England in the eighteenth century. In other countries, especially the US and elsewhere in Europe, the details of their history of procurement are different, but the methods of tendering, contracting and the professions that emerged in England around the turn of the eighteenth century at the beginning of the industrial revolution became the basis of the modern system of construction procurement.

Early Procurement Methods

Through the Early Modern period in Europe, until the seventeenth century, the usual way of getting building work done was to employ craftsmen directly. If the client did not supervise the work himself, an agent would be appointed, and craftsmen were employed at daily wage rates set by medieval craft guilds that guaranteed quality and were responsible for training. The trades and guilds were based on the materials used, such as thatchers, carpenters and stonemasons, and had an apprentice-journeyman-master structure that set standards of work enforced by the courts run by their Companies.[64] This had been the normal way of building for centuries, although even then there were other methods of employment and payment.

While the common system was direct employment by the client at day rates, sometimes work was done under a system known as 'measure and value', where work done in the different trades was valued based on an agreed set of prices and wages rates. Master craftsmen would make an agreement (similar in effect but not a contract) for building and employ other craftsmen and labourers to do the work, called task work. The payment was still usually on a measure and value basis not for an agreed price, although there was some lump sum work.[65] As use of this method of 'contracting' increased use of the direct labour system declined, although wages and prices were still largely regulated by the guilds. Measuring was usually done on completion and had its own issues, mainly with delay, disputed work and associated litigation.

A third way of contracting was known as in grosso (by the great), meaning for an agreed, fixed sum. Harvey found examples from the fifteenth century where what today would be called a lump sum contract was used to deliver projects. An important difference was that they involved separate agreements with the various tradesmen, however in this case the price was agreed in advance instead of work getting valued when done. This form of contracting was strongly opposed by many clients and craftsmen, who felt that the in grosso formula would produce poor quality and high prices, and sureties were often demanded to guarantee completion if the contractor did not complete the work as agreed.[66] The method of contracting 'by the great' led to the end of regulated prices and slowly undermined guilds, eventually leading to the end of their role as monopoly suppliers of craftsmen.

There are also early examples of fixed-price contracts, typically with the separate tradesmen involved on a particular project. These were quite rare. It was not unknown, however, for a single person to carry out a complete project through subcontracts or direct employment, which led eventually to the modern contracting system. The building of a jail in York in 1377 used a fixed price and a contractor with agreed stage payments,[67] a good example of the continuity of many of the methods of procurement over the centuries.

Christopher Wren, in a 1681 letter to the Bishop of Oxford, explained that there were three ways of getting a job done and the problems with each one: working by the day, by measure, and by the great. His preference was to work by measure, although good measurers were hard to find, and he argued that contractors employed by the great who were not familiar with the tasks 'doe often injure themselves, and … shuffle and sligh the worke to save themselves'.[68] In the building of St Paul's after the 1666 Great Fire of London, Wren was responsible for both supervision of the work as well as the design, He employed different sets of craftsmen under these three different types of contracts to do the work, but most of the work used the method of separate contracts with different trades, a form of contracting by the great. This spread the risk and became the most common form of contracting.

By the end of the seventeenth century some procurement was being done through a building agreement, usually for speculative property developers, between the client or developer and the builder. Agreements were legally binding documents that specified the structure or structures, their size and materials, payments and any financial penalties incurred after the agreed time. These were described as 'fairly comprehensive and sophisticated'.[69] They contained arbitration clauses, and resort to the courts only happened when arbitration failed. The main problem was quality control, and common phrases like 'well made' or 'good work' made disputes over quality common.

The growing role of property developers and speculative builders is an important element in the development of London's building industry. Some of these were lords (Bedford, Southampton, St. Albans) but many were not (Babon, Cubitt, Bond). Literate and numerate – skills essential for estimating – some master craftsmen became master builders. Usually bricklayers or carpenters, these men would build single houses for the speculative market and employ or contract with tradesmen. They used surveyors to mark out building sites and measure completed work to settle contracts. A wave of house building by these small builders and developers started in the mid- seventeenth century, which over the next 50 years led to large property developments with a general contractor responsible for many buildings on an estate. These ambitious craftsmen and opportunistic developers were not deterred by city officials or prevented from building by royal decrees or church interests, unlike the situation commonly found in other European cities: 'London is above all a metropolis of merchandise. The basis of its building industry is the trade cycle rather than the ambitions and policies of rulers and administrators. The land speculator and adventuring builder have contributed more to the character of the Georgian city than the minister of the Crown ... or the monarch'.[70] This was a new system of production, with a general contractor who had overall responsibility for building work done and a developer that financed the project and acted as an intermediary between the eventual consumer and the producer.

Guilds, Trades and Professions

At the turn of the century construction was undergoing structural changes to the organization of work. For some hundreds of years before 1800 building work had been done by independent craftsmen belonging to guilds, or Companies, who usually worked directly for a client. Guilds were effectively local monopolies, an alliance between business and (typically) the crown or noble that provided protection, restricted competition and reduced threats from innovators, competitors and new entrants in return for tax revenues. Guilds protected members' interests, a privileged few, and were exclusive organizations for relatively well-off, middle-class men with the purpose of enriching their members at the expense of consumers and non-members. They were small relative to the markets they monopolized, so the costs were widespread, and they were small relative to the wider labour market:

> Occupational guilds in medieval and early modern Europe offered an effective institutional mechanism whereby two powerful groups, guild members and political elites, could collaborate in capturing a larger slice of the economic pie and redistributing it to themselves at the expense of the rest of the economy. Guilds provided an organizational mechanism for groups of businessmen to negotiate with political elites for exclusive legal privileges that allowed them to reap monopoly rents. Guild members then used their guilds to redirect a share of these rents to political elites in return for support and enforcement. In short, guilds enabled their members and political elites to negotiate a way of extracting rents in the manufacturing and commercial sectors, rents that neither party could have extracted on its own.[71]

Guild members thought there was a limited lump of labour to go around, and innovations that produced more output from existing inputs would flood markets, and therefore opposed innovations that threatened their rents in this zero-sum world. They were against new devices and products, forbade their members to adopt new processes, blocked imports and boycotted foreign products. The way they operated was to set both price ceilings for raw materials and piece-rate ceilings for subcontractors. Guilds could also impose 'monopoly contracting', outlawing sales and purchases by individual craftsmen and merchants. These created a rigid regime of collective prices and quotas that removed craftsmen's incentives to do better work and merchant's incentives to experiment. To defend their monopoly prices, guilds used quality regulations to prevent their own members from producing higher quality products, regularly inspected work and could demolish work considered unsatisfactory.[72] Innovators could be punished or expelled for disrupting the order of things.[73]

The end of the eighteenth century, however, was the beginning of a transition from these old, established ways to what eventually became known as the 'modern system' of contract labour and measuring to determine costs. This was not some linear, steady progress, but an overlapping of the old procurement methods with the new system, as both continued to be widely used:

> The guild system had broken down well before the late seventeenth century and certainly in London any remaining vestiges of power that the Companies had were annihilated by the legislation following the Fire which allowed 'foreigners' from outside the city to work within its boundaries. The building industry might still be organized around separate trades, however the relationships between these different crafts and the methods of contracting were undergoing a profound transformation.[74]

Importantly, as part of this transition to an industry with more of the characteristics seen today, the different construction professions began to form, as experienced tradesmen started to specialise in various aspects of building and construction and professional architects and engineers appeared. The emergence of the roles of architect, engineer, surveyor and contractor began during the rebuilding of London after the Great Fire of 1666 under Christopher Wren and Robert Hooke.[75] A hundred years later, toward the end of the eighteenth century, the general contractor had arrived as a new type of firm, responsible for organising the building process and employing craftsmen to undertake work directly or as subcontractors.[76]

THE TRANSITION FROM THE CRAFT SYSTEM

Until the seventeenth century building had typically been done by craftsmen working directly for a client. Well into the eighteenth century clients would buy the materials and pay for labour only, in other cases the craftsmen would supply their own material and agree to a price for the whole job beforehand or work on a value and measure basis. However, by the end of the 1700s the measure and value method was being replaced by contracting with one person to undertake a project for an agreed price, although still disliked by both tradesmen and employers. In the early nineteenth century procurement through contracting became the norm, and it has remained so since.

From the late 1780s there were regulations requiring British government departments to use the contract in gross, but this was generally ignored. Administrators in the Office of Works and Public Buildings were typically against contracting for a fixed sum as well as contracting for the whole project, and they carried on with the old way of employing separate trades until a reorganization in 1832 forced a transition from separate contracting with unit prices to general contracting for a lump sum.[77]

There were four types of building firms found during the late eighteenth and early nineteenth centuries[78]. The first were master craftsmen, employing journeymen and apprentices and working with their own trade (the traditional medieval system). The second were master craftsmen who contracted for a whole building but then contracted with other master craftsmen for work outside their own trade. This was largely a barter system known as 'blood for blood', however this form of co-operative contracting had largely disappeared by the middle of the 1800s.

The third type of firm were builders, often architects, who completed buildings by contracting with master craftsmen in each trade. During this transition from direct labour to a contracting system in Satoh's description: 'The architect as the agent of the building owner assumed the undifferentiated duties of designer and supervisor on the one hand and construction manager on the other' (1995: 16). These architect-contractors on large projects were a passing phenomenon, although they continued to do minor works throughout the nineteenth century. With the establishment of the Royal Institute of British Architects in 1834 the distinction between builders and architects was made clear and became widely accepted by the profession. One consequence of this was the rise of merchant builders, who had access to both materials (their form of business) and capital. Some of these were speculative developers and some became large scale contractors.

Type four firms were the master builders, who permanently employed their workers and were responsible for the whole project, often but not always won through a competitive tender. These firms were the original general contractors, and from the early 1800s contracting for the whole project at a fixed sum became widely adopted, first in the private sector with the public sector following the trend. Some of the London-based firms doing major projects became very large, employing several thousand men each by mid-century.

These changes led to the disappearance of the craftsmen builder in London over the last two decades of the eighteenth century, and by the early part of the nineteenth century the new system of contracting was well-established.[79] This new contracting system, with one firm doing the work for a fixed price, was opposed by many clients, architects, tradesmen, and small contractors. There were also problems of fraud and business failure associated with fixed price contracting. These objections are familiar today, and many of the comments reported then are still found in the succession of UK industry reports Murray and Langford detailed nearly 200 years later.[80]

The new competitive system was fatal for the guilds and their long-standing traditions and practices. The most bitter opposition, however, which continued throughout the nineteenth century, came from tradesmen and building workers. Opposition to general contracting was driven by a combination of increasing use of machinery, both on-site and in the workshop, and the increase in the length of the working day to around 12 hours. Indeed, a hundred years later at the close of the nineteenth century, trade unionists were still arguing about craft rights and opposing building work done with contracting by one person or firm. Fear of technological unemployment was another important factor.

The smaller contractors, master craftsmen and small builders also opposed the new system because they often lost money by bidding too low for contracts. Many of these became employees of the large contractors, a significant loss of social status. Despite the opposition, the new procurement system spread rapidly during the early years of the nineteenth century and became a normal way of doing business. By the 1830s both private and public clients had come to believe the design-bid-build system was the best way to contain costs within estimates and to get value for money. These remain the primary concern of clients to this day.

WESTMINSTER PALACE

The construction of Westminster Palace (the Houses of Parliament in London) from 1837 was one of the first major buildings to be done using the 'modern system' of procurement and contracting, and the example of the building of Westminster is instructive.[81] The project used new engineering techniques and machinery, the skills of hundreds of traditional craftsmen and a huge work force managed by some of the largest contractors. There were between 400 and 1,400 men employed on the project at any one time; in 1848, 776 were on site, 120 in quarries and 335 in the contractor's workshop. The project was expected to take six years and cost £700,000, but actually took almost 30 years and cost over £2 million (well over £500 million or USD$1 billion in today's money).

There were disputes between the architect who won the design competition, Charles Barry, and pretty much everyone else involved, starting before construction began. There were arguments over the initial design, over the estimates and the architect's fees, and disputes over contract prices and problems with the materials. The designer of the heating and ventilation system fell out with Barry and the two became enemies. The masons went on strike for 30 weeks after a foreman swore at them. The size of the project created more problems in logistics and along the supply chain.

The project was carried out through a series of successive contracts awarded through competitive tender or by recommendation by Barry. The first two contracts were let by the government department in charge. The third contract was put out for tender to eight firms recommended by Barry, and was won by one of the largest London contractors, Grissell and Peto, who were then given the following four contracts by negotiation without further competition. In the third contract the prices were set by the builder and agreed to by the architect, however in the fourth contract prices were determined by the government department and set lower than prices arrived at through competition. The assumption was that the new industrial technology appearing on building sites, such as mechanical scaffolding and steam powered hoists and pumps, would reduce the costs. These contracts had detailed specifications on all aspects of the work and were priced in a form recognizably similar to a modern bill of quantities. The work of each trade was specified separately. The contractor later renegotiated a new set of prices as these were profitless contracts otherwise.

The 1844 contract for 'finishings' was on a unit price basis, reduced by 22 percent by the Office of Works and Public Buildings on the basis that the new machinery coming into use would save time and labour. As it turned out much of the carpentry work, although initially carved by new machines, had to be completed by hand at significant extra cost. Grissell terminated the contract in 1845 and the complex negotiations that followed went on over many years. Building agreements in those days did not include provisions for claims and variations.

The Modern System

By the beginning of the 1800s builders were becoming recognised as an occupation, and as their firms developed they often specialised in certain types construction, such as civil engineering during the boom in railway building later in the century. Others, such as William Cubitt, became developers, building housing to meet demand from the rapid growth of cities. By the middle of the nineteenth century large contracting businesses had taken on the form that in many ways we still see today, and procurement and contracting was using the same, or a recognizably similar, system.[82]

There were two key characteristics of this new procurement system. First was the use of detailed drawings and design, completed before the work began. The second was the preparation of cost estimates for the project, on the basis of the design drawings. The two significant outcomes of these characteristics, that became the foundations of the modern system, were the shift to competitive tendering and the growth of the professions. Buildings began to be done with detailed drawings from the architect and a bill of quantities (BQ) with full estimates based on them, in contrast to the system of measure and value where costs had been determined on completion by a measurer, originally a tradesman, which over time became a specialised task. As the new method of procurement and contracting appeared measurers became more important, as prices had to be agreed between the architect, client and builder before work began, thus measurers became quantity surveyors.

A series of government commissions and inquiries on building procurement produced reports that sometimes, but not always, favoured competitive tendering.[83] Nevertheless, as the system became more widespread government departments came round to the idea that it was the best way of obtaining value for money. By the middle of the nineteenth century competitive tendering on the basis of design and price specification had become the usual practice. The new profession of quantity surveying was therefore an essential element in the new system.[84] The procurement method where clients invite tenders on the basis of completed designs made it necessary for the client to know whether the tender prices were reasonable, so the design had to be quantified in terms of materials and labour, and then priced. As explained by construction economist Jim Meikle:

This new process brought about the need for contractors to not only have drawn information but also a clear description of what was required in terms of the quantity of work. Contractors quickly realised that it did not make much sense for all tenderers on a job to produce their own quantities so they got together and appointed one person or firm to undertake the measurement: the quantity surveyor. The list of measurements was called a bill of quantities.[85]

While bills of quantities dated back to the middle of the 1700s, an 1828 parliamentary committee investigating the Office of Works and Public Buildings found the practice had become well-established. Bills of quantities become fundamental to the contracting system in Britain, and later became part of the required contractual documentation. It is worth noting that other European countries did not find the detailed bill as essential.

Not all of the projects using the new system were won through competitive tendering, often contracts were negotiated with firms familiar to the client or architect. In these early days of fixed-price contracting, the idea of competition was controversial. Much of the opposition to the contracting system was really opposition to competitive tendering, rather than to the idea of a single contract with an agreed sum. Competition was believed to lead to lower standards as contractors would bid low to win work and could not possibly match those prices without reducing standards.

It was also feared the contractors would abuse the system through collusion on bid prices or corrupt practices. Fears that have been justified more than once, as cases of bid rigging, cover pricing, successful bidders payment of fees to unsuccessful tenderers, and market sharing agreements, which were all reported to occur then and are also found today. The protection against such practices was initially based on the idea of only employing 'respectable' builders, and the idea of respectability was seen as protection against the consequences of competition.[86] Rather than reliance on builders' respectability, clients' best protection against abuse came to be seen as careful pricing of specifications and close supervision of the work by the architect: 'by the end of the nineteenth century the relative positions of builders and architects had reversed. Builders were now instrumental to the wishes of the architect, instead of themselves taking charge.'[87]

One key element of the builder's respectability was possession of sufficient capital and employees to carry out a job without subcontracting, which was regarded as a dubious practice. Clients wished to avoid subcontracting, so it was often done secretly. The original large contractors employed craftsmen in every trade and were expected to complete most of a project under their own management. However, the advantages of subcontracting led to its widespread adoption by the middle of the nineteenth century. The advantages then still exist today, such as flexibility of employment, managing risk and liability, and specialisation, which was important as the development of new materials and new components required new skills (for example patent glazing, iron and steel frames, gas, and later electrical and lighting).

This was also a time a rapid technological innovation and development, both by and for contractors.[88] Technical advances occurred in stone, wood, bricks, components, pumps and lifting machinery. Like other industries the widespread availability of steam power was transformational in the application of new machinery in the contractors' workshops, and the use of mechanization on building sites slowly increased. There was also an ongoing transfer of site work into the workshops. For the largest firms these were huge, William Cubitt (contractor brother of property developer Thomas Cubitt) had 25 acres on the Isle of Dogs in 1845, complete with wharves, sawmills, cement kilns, an iron foundry, brickfields, a pottery and so on, linked by an internal railway and employing about 800 men.

In 1818 the Institute of Civil Engineers was founded and in 1834 the London Builders' Society, partly in response to the rise of the labour movement as the influence of the guilds declined with competitive tendering. By coincidence this was the same year as the founding of the Royal Institute of British Architects (RIBA). The main purpose of the Builders' Society, however, was to hold together builders 'who being asked to tender on a specification that did not contain an arbitration clause had all declined. The arbitration clause seems to have indicated what was to be done in the case of controversy between owner and builder'.[89] These arbitration clauses were the source of many disputes and conflicts between contractors and architects, and led eventually to the Conditions of Contract agreement.

Architects strongly favoured traditional contracts for price over contracting in gross, at a fixed sum for the whole project. This avoided 'the tedium of preparing the correct drawings and specifications beforehand',[90] instead of preparing designs and giving directions as the work progressed as they used to do. However, as conflicts between owners and contractors became more common, and intense, under the new competitive system, architects realized the importance of the role of the project superintendent. With the central role of the architect as client representative becoming established, the RIBA wanted to ensure architects were seen as acting on behalf of their clients. Concerned about potential conflict of interests and protecting clients the RIBA, in 1887, prohibited members from getting involved and profiting from the organization of building work. Architects in the US had the same prohibition, according to Howard Davis: 'by the 1890s in New York, the building culture had taken on essentially the form it has today. The architectural profession was emerging as an independent set of institutions from those of builders, developers, banks and regulators – each of which had its own identity and asserted its own authority.'[91]

CONTRACTS, CONFLICT AND RIVALRY

Under the craft system the contract by measure was used, agreements were made with individual craftsmen and paid at standard rates. Often these agreements were verbal, based on a shared understanding of what was required and expected. This form of contract involves a lot of implicit knowledge and understanding shared between the participants of a building's type, use and construction. Early written contracts had no drawings or plans, and few details on materials and fittings. While there might be general references to a 'well built house', the number of windows, 'proper hinges' or 'burnt brick', the master builder or craftsman had responsibility for design decisions such as where to put the windows, stairs, fireplace and chimney. The cost was finalised when the building was complete.

The contract in gross and competitive bidding placed a general contractor between craftsmen and the client. To prepare a bid the building must be specified and measured from the drawings to arrive at a total cost before work starts. As competitive bidding became more common over the nineteenth century, contracts became more and more explicit in the terms used and specifications given. Architectural historian Howard Davis used two dozen London contracts from the middle of the seventeenth to the late nineteenth century to show how three fundamental changes developed over that time:

1. The inclusion of drawings and reference to them. A 1668 contract for a five story building referred to the drawing once, the 1774 contract had plans and elevation with features specified, by 1822 a large building had 33 drawings on commencement and more expected.
2. The quantity, nature and precision of specifications and the character of language. Early contracts specified brick and mortar, later ones identified different parts of the brickwork. By 1872 details on facings, plastering, hardware and finishing were included. 'Not only is there increased detail ... adjectives that call for subjective interpretation disappear', replaced by descriptions with 'considerably more detail'.
3. Professional responsibility for the building. By the end of nineteenth century the architect was responsible for interpretation and compliance with the contract. The trend had been reinforced by the growing legal importance of specification in the contract, which also gradually reduced the role of builders' discretion and judgement. The increase in specification was linked to the development of building regulation. The first use of a building code was for fire safety in the rebuilding of London in the 1670s after the Great Fire, followed by more (six over a hundred year period) on height, proportions, party walls and quality.[92]

The idea of the contract is to make clear what the obligations of each party has, but no one has ever devised a contract that eliminates all possibility of disputes over interpretation and performance. As the contracting system developed it was the architect who came to determine the conditions under which work was let and was responsible for resolving disputes. Under the modern system these contracts gave architects a unilateral power to determine payment to contractors, which was sometimes abused to benefit clients, and was the source of bitter complaints from contractors. Note that not all projects were done by contract. Successful tradesmen who did their own small scale developments and general contractors with large speculative developments did not use them because they controlled their sites directly

In 1870 the terms of a document called the Heads of Conditions of Builders Contracts was agreed between architects (RIBA) and the Builders' Society This established the basic outline and principles of the standard building contract which could then be varied to particular circumstances. It addressed the concern of builders who felt that previous contracts made no provision for variations in materials prices or the cost of extra work. Bills of quantities were introduced as part of the contract in 1902, after many revisions in the meantime, and this remained the basic form until 1931 when the Joint Contract Tribunal was set up and the standard forms of contract came into use.[93] These are still the basis of the majority of building contracts in the UK today.

RIVALRY BETWEEN ARCHITECTS AND ENGINEERS

As well as the conflicts between architects and builders, there was considerable rivalry between architects and engineers. This began in the early nineteenth century as the pace of technological innovation increased and new materials arrived – iron, then steel, followed by reinforced concrete at the end of the century - and mechanical engineering emerged (a British specialization) with railways and the new steam-powered machines and equipment. Architects knew little about these innovations and left them to engineers. By 1800 architecture and engineering had become separate professions with separate training. Architecture retained elements of the apprenticeship structure as they worked under older architects and in studied in schools of architecture, while engineers went to engineering faculties. The antagonism found in the UK between architects and engineers in the early nineteenth century was also present in America, described as a 'great schism' that developed as architects struggled to master the requirements of new forms of building and new materials and the mutual contempt that grew between them and the new profession of engineers.[94]

The disengagement of architects and design from building and construction occurred at a time when engineers were also focusing on design rather than delivery, due to increasing specialisation and differentiation between different types of engineers (such as mechanical, structural, civil, electrical and sewerage). Both architects and engineers neglected estimating, which was left to the new professions of quantity surveyors and cost engineers.

Thus, each of the professions developed their own language, skill sets and cultures, and shared a sense of superiority over builders and contractors. At a time when science was transforming construction materials, and mechanization was industrialising site work and components, the disassociation of participants in the building process and a general lack of interest in technology were already becoming issues. Architectural historian Andrew Rabenek argued:

> Following early establishment of the basic contractual transaction model, modern building roles and practices matured in the long period up to the First World War. Professional and craft associations were established ... property and construction law were accumulated ... Schools and training programmes were set up ... secondary industries supporting building began to industrialize ... Despite this commercial, social and institutional development, building itself remained largely traditional in its technology and craft practices. Tradition was experience crystallized into rules of practice that could be taught and passed on by master to apprentice. Only in structural engineering was the application of science apparent before the twentieth century.[95]

That was about the UK, but this disengagement was also seen in the US. The first school of engineering was founded in New York in 1824, and in architecture at the University of Illinois in 1867. The American Society of Civil Engineers and Architects was founded in 1852 in New York, but in 1857 the architects split to form the American Institute of Architects (AIA) due to architects' opposition to engineers working as "package dealers", i.e. undertaking design-build contracts. Architects were prohibited from design-build work and any financial relationship with a contractor, manufacturer, or supplier, part of the AIA Code of Ethics until 1978. After the formation of the AIA, engineers became the American Society of Civil Engineers then, as they became more specialized, formed two additional organizations: the American Consulting Engineers Council and the National Society of Professional Engineers. These organizations published their own standard form contracts. The first AIA design-build contracts were published in 1985.[96] In the UK, the Joint Contracts Tribunal in 1981 published their Standard Form of Building Contract with

Contractor's Design for design-build projects. As in the US, this required changes in the codes of practice for British architects and chartered surveyors, to allow members to work in design and build firms.

THE RISE OF THE CONTRACTOR

By the end of the nineteenth century in the UK there was a fully developed procurement and contracting system with practices well understood by all the parties concerned, and this system continued, with its essential characteristics unchanged, into the twentieth century.[97] Nevertheless, it also continued to generate controversy and conflict, and an increasingly litigious industry. In the first half of the twentieth century the modern system was refined and further developed. Construction historian Marian Bowley outlined four ways of contracting as characteristic of the period between 1944 and the 1960s. First was selective tendering, where only contractors known to be capable are invited to tender. Second were negotiated contracts, often used by local housing authorities to bring the contractor in at an earlier stage. Third was serial contracts, with contractors having successfully completed one project were re-engaged on later ones. Fourth were package deals or design and build, used for the UK's mass housing programs in the 1960s.

None of these were really new. All of them had been used before in various forms and they have all reappeared, often rebadged. An ongoing proliferation of contract forms to overcome inadequacies of the traditional system have led to an ever-expanding variety of contracts and procurement systems to choose from. How effective this has been a topic for debate. On construction procurement Bowley said 'It is difficult to see how any system more wasteful of technical knowledge, intellectual ability and practical and organising experience could have been invented'.[98] While it is hard to disagree with the sentiment, it seems to overlook the evolution of procurement methods as new versions and new types of contracts developed as a response to problems and issues found in the existing procedures. This is what led to the use of design and build contracts and the increasing importance of contractors in the second half of the twentieth century.

By the middle of the twentieth century package deals or design-build contracts were becoming common for large projects, and by the end of the century were in widespread use for all types of projects. For many clients this was a way to offload the risks associated with their project and a solution to the complexity and difficulty of managing a number of separate contracts, by giving responsibility for both to one contractor. This contractor led form of project organization was significantly different to the traditional designer led form with its design-bid-build structure, and the roles of both clients and architects were reduced.[99] As the top panel in Figure 5 demonstrates, on a designer led project the client and design professions are the leading participants in the early stages of a project, with contractors responsible for site work. On a contractor led project, however, the contractor is the leading participant from project inception.

Large contractors delivering major projects ended up at the core of construction at the end of the twentieth century.[100] By this stage the organization of the production system had a clear outline, and a very clear structure, for bringing together the products, suppliers and materials needed for building and engineering projects, and had stabilised around particular forms of procuring, financing and managing those projects and their political, organizational, and technical factors.[101] Construction was described by Dubois and Gadde as a 'loosely coupled system',[102] where temporary project teams come and go, combining members from a broad range of firms and industries as required, to arrive at viable solutions. These project teams combine many specialisations, and how they are organised and function is to a large degree determined by the type of contract used. As the saying has it 'the rules are the game'.

Figure 5. Organization of designer and contractor led projects

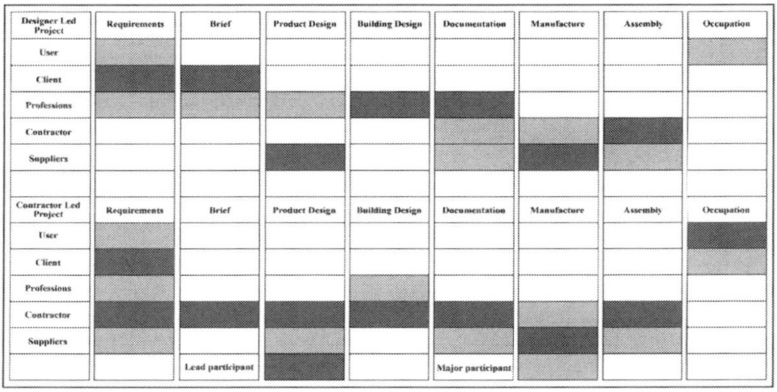

What this history shows is how, over a period of 200 years, a system of procurement and contracting based on measurement and specification, replaced the older systems of direct employment of craftsmen at day rates and measure and value. As this new system was developed and maintained it had great continuity, which is an important element in understanding how difficult innovation in procurement actually is. The fundamental changes seen since the modern system of procurement came into widespread use in the early nineteenth century did not replace or overcome issues between clients, consultants and contractors. Procurement methods evolved slowly, in response to problems and issues with the methods in use and to changes in both the organization of work and the structure of society.

During the nineteenth century general contractors, often winning projects in competitive tenders, became responsible for organising projects and employing workers. While there were recognizable elements of the old craft system, the building industry was becoming a more complex and confusing conglomeration of businesses, professions and individuals. Many of the characteristics of the industry that emerged during the transition from craft production to the modern system during the industrial revolution are still part of the building and construction industry today, as was seen in the example of the building of Westminster Palace. The same can be said for many of the issues found in procurement and contracting for Westminster.[103]

Despite being constantly criticised and modified around the edges, procurement at the end of the twentieth century is still found have issues.[104] Inquiries and reports in many countries like the UK and Australia concluded government intervention and/or contractual reform was needed.[105] In project manager Peter Morris' view:

> It is like a game. There are rewards and penalties, rules and roles. Some cheat, or at least take advantage, where others wouldn't. Some play the game straight and true, others are always looking for an angle to make another dollar or two. Or three. Contracts describe what is to be provided under what conditions. Some people put the contract in a desk drawer and forget about it, others use it as a means of extracting increased payments. The contract sets the rules but it is the individual who decides how play will be conducted.[106]

These comments echo those made in the mid-nineteenth century. Satoh's history of the first industrial revolution and construction[107] closes with a series of quotes from opponents of the modern system. These included: poor quality work due to low price bidding, or subcontracting; builders undercutting each other to win work; the lack of provision for variations in fixed price contracts; unjustified claims by contractors; arbitrary decisions by superintendents and architects; non-payment by clients; and collusion between contractors. To address these issues the contracting system incorporated increasingly detailed drawings and specifications, and a schedule of prices was often attached for claims and variations. The unilateral nature of the contract led to the drafting of the Conditions of Contract, which were revised over time.

This history is focused on England and its development in London, because that is where many of the first modern projects were built. Competitive tendering, enforceable contracts, subcontracting, surveying and measurement of costs with a BQ are now widespread, but these all came with the modern system that developed in the UK, a significant innovation. Other countries have different histories that emphasise the role of engineers rather than architects, but this modern system of procurement was the foundation upon which they also were built.

The traditional craft system of production with its guilds and their companies was replaced by a new system with tradesmen and contractors, blown away in a 'gale of creative destruction.' The emergence of the modern system of procurement and organization of production then created the institutional and contractual framework that binds engineers, architects, manufacturers, contractors and subcontractors together in construction of the built environment. This system of production developed over the nineteenth century during the first industrial revolution as the various elements of the modern industry came together, pulled along by ever larger and more complex projects building canals, roads, bridges and tunnels, railways, factories, offices and housing.

Figure 6. Construction of the built environment

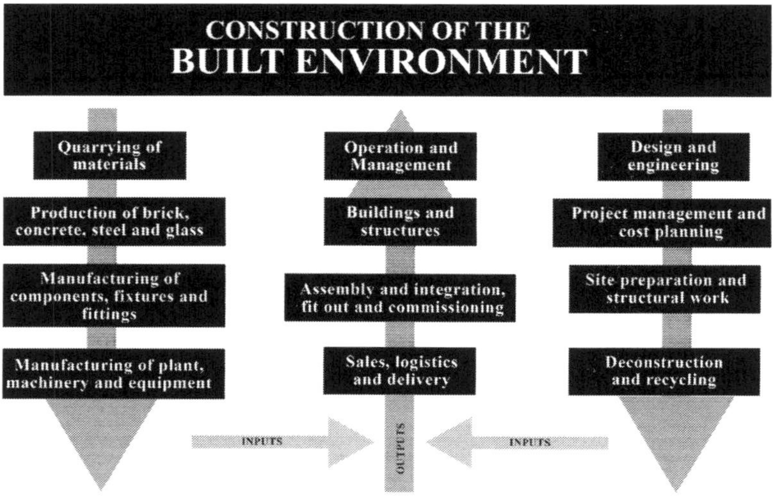

3 The First Industrial Revolution and the Industry Lifecycle

Each industrializing society develops its own combination of elements to fit its traditions, possibilities and circumstances
David Landes[108]

MODERN CONSTRUCTION EMERGED during the first industrial revolution in the early nineteenth century, followed by a period of rapid, disruptive technological development not unlike the present one in the late nineteenth century. Between 1860 and 1900 building and construction was restructured as an industry by the rise of large, international contractors, and project management and delivery was reorganised around steam powered machinery and equipment. Major projects like the Suez Canal, railways, tunnels and the new factories for mass production were typically built by new, global European contractors employing workers from around the world on their projects. These projects also required a new organizational form that integrated components, systems and processes. In materials, with steam powered industrialization and mechanization the disruptive new technologies of steel, glass and concrete came together in the last decades of the century, causing fundamental changes to both processes and products.

Why would the experience of the industry over 100 years ago be relevant today? There are two parts to the answer. The first is that the second half of the nineteenth century is the only other period of disruptive change similar to the present available for comparison. The second is that the effects of technological change on industry structure and performance might plausibly again be in the same key areas as the organization of projects, industrialization of production and the mechanization of processes, but in the twenty-first century these effects will be heightened and quickened by the network effects associated with digital platforms and AI.

The chapter first explores engineering historian Tom Peters' three 'dimensions of development.' This is followed by analysis of the role of invention and innovation in construction and historian Thomas Hughes' model of the industry life cycle, where he identified seven phases in the life cycle of a technology. The following section focuses on invention and development in the first four of the seven phases where the effect of previous GPTs and their impact on building and construction is discussed, with the mechanization of construction with steam power in the nineteenth century used as an example. The following section covers the next three life cycle phases of growth, competition and consolidation and discusses innovation in production and organization. The significant role of innovation in materials is also discussed, with incremental innovation in concrete used as an example of how new materials affect the organization of construction.

Dimensions of Development

The various elements of modern construction came together over the nineteenth century, pulled along by ever larger and more complex projects building canals, roads, bridges and tunnels, railways, factories, offices and mass housing. During the 1800s the world was urbanising as population rapidly increased and major cities attracted migrants and businesses. In the second half of the century heavy industry and manufacturing spread around the world, from England and Western Europe to America then Japan. New industries needed new types of buildings, typically larger, higher and stronger than traditional methods and materials could provide. Demand for new types of buildings was important in stimulating technical and organizational innovation, and construction was an adopter of innovations from other industries rather than a source of innovation, and innovation in basic materials played an important role.[109] For example innovation in materials such as concrete, bricks and glass played an important role as the development of new tools and techniques sought to take advantage of them. In recent years this has been seen again, with early adopters of BIM benefiting as competitors were catching up.

As a technology-based system of production, inventions and innovations by firms that affect the onsite work done can be included in discussion of the effects of new technology on production of the built environment.

For example, hydraulic lifts in the 1880s, and high-strength concrete in the 1980s. In fact, the period between 1860 and 1900 was one of great disruption due to product innovation in building materials, and provides an example of an industry changing to a new organization of production in response to technological change. Over that time, there were technological shocks to the existing building and construction industry as the new materials of iron, glass and concrete opened up opportunity and possibility for designers, in both how and what was built.

Iron and steel separated the building frame from the envelope between the Crystal Palace in 1851 and the rebuilding of Chicago after the Great Fire of 1871, and with the separation of the frame from the envelope came mass produced infill materials and facades to replace load-bearing construction.[110] Then the combination of steel and concrete made possible the development of reinforced concrete and steel skeleton structures. Both 'building art and the art of building' were transformed, not once but several times, over these decades as the new methods of industrialized building using steam powered equipment and iron, steel and reinforced concrete were refined.[111]

The adoption of steam power is an important precedent, a previous experience of technological disruption leading to a restructuring of construction. In *Building the Nineteenth Century* engineering historian Tom Peters called the three areas of construction that were transformed in the nineteenth century 'dimensions of development'. These were industrialization, mechanization and organization:

1. Industrialization of production methods with standardisation of components and mechanised mass production, and the development of new materials like steel, plate glass and plastics. This led to a new design aesthetic, with more modular components and internal services, and separated the envelope from the structure. The infrastructure of materials suppliers and equipment producers developed, and scientific R&D joined construction's traditional trial and error approach to problem solving.
2. Mechanization of work based on steam power, with cranes, shovels and excavators becoming common by the mid-1800s.

This in turn led to a reorganization of project management, with the new form based around logistics and site coordination to maximise efficiency of the machinery and equipment.
3. Organization of modern construction was emerging. Large general contractors had appeared by the 1820s and were undertaking projects on a fixed-price contract often won through competitive bidding. This system of procurement was supported by the new professions of architects, engineers and quantity surveyors, which had emerged during the eighteenth century and were institutionalizing in the early nineteenth century.

An example of the interaction of these dimensions is the Suez Canal, the largest project ever undertaken when started in 1862. The first phase began with a French contractor and used up to 40,000 Egyptian serf labourers to dig 75 kilometres, but in 1863 the workers were withdrawn by the Egyptian Government, the chief contractor resigned, and the project was reorganized. By 1866 French industry had produced the steam powered excavators, conveyors, cranes, tractors, railways, dredges and barges the new (English) contractor required, and by 1867 the site was fully mechanized. Between 1866 and 1869 another 85 kilometres were dug, 50 million cubic meters of earth moved, and a novel form of site organization based on 'planned processes' began to emerge. In the year before opening in 1869 there were 7,000 labourers and 6,900 European workers on the project.[112]

Fifteen years later construction of the Panama Canal started, and Peters argues the modern form of site organization and project management emerged when the US military took over in 1904, again after the first attempt at the project failed. Peters concludes 'at Suez, the most innovative thinking had gone into design and manufacture of machinery. At Panama, Goethals and his team emphasised the design of integrated systems and processes.'[113]

At Suez the relationship between technological change, conceptual thinking and organizational form is clear. While the striking thing is the interrelationship of these three aspects, the driver of these changes is technology, or more precisely a new technology that fundamentally changed existing industry practices and delivered a shock to the existing system of production. With the advent of iron-framed and reinforced concrete buildings the construction industry had to not only master the use of these new materials, but also develop the project management skills the new technology required. That organizational change, in turn, was based on the deeper change in the way of thinking about the world that was fundamental to the industrial revolution and the invention of the scientific method.

Another example that highlights industrialized production and the effectiveness of combining innovations was machine made nails.[114] With industrialized production, prices of manufactured goods decline over time with new production techniques and increasing economies of scale and scope.[115] Over time those cheaper prices allow new technologies to spread and find new uses. Originally nails, like everything else, were hand-made and more expensive than screws, but by 1828 the cost was down to 8c per pound (nearly half a kilo) and in 1842 to 3c. Dimensioned lumber and cheap machine-made nails made possible a whole new order of speed and economy in wood framing. Combining these two innovations a new system of building known as the 'balloon frame' came out of Chicago in the 1840s, and with nailed light timber frames two people could do the work of twenty using traditional methods.[116]

Skilled labour in the US was then scarce and expensive, so this tenfold increase in productivity from two relatively simple innovations had a major impact. Balloon frames were sold in catalogues in many styles and were used to build the new railway towns and suburban housing that spread across America over the following decades.[117] This shows the importance of a combination of new innovations within a technological system, which is often more significant than the individual new technologies themselves.

A current example of a combination of technologies is smart helmets, which use augmented reality (AR) to combine BIM and location data. Trimble, a major construction software provider, is using Microsoft's HoloLens onsite in a smart helmet called TrimbleConnect, which provides a mix of the real with AR. The Trimble helmets join the Daqri smart helmet already available, a more expensive option with a built-in expert system.[118] Combinations of BIM and offsite production methods are being developed by housing manufacturers and by firms like Autodesk, Mighty Buildings and Project Frog. Engineered wood factories prefabricate cross laminated timber (CLT) components combining BIM and computer numerically controlled machines. CLT was developed in the early 1990s in Austria and in 2015 was incorporated into the International Building Code.

Invention, Innovation and the Industry Life-cycle

Industry life-cycles are drawn as an S-shaped curve, similar to a product life cycle, often with the five stages of development-launch-refinement-growth-decline. Industries start small and fragmented with many new entrants, grow quickly when products become widely accepted, and then consolidate over time around core products and processes as growth slows, eventually turning into a 'sunset' industry. An engineer and historian of technology, Thomas Hughes' identified seven stages of industry growth from inception to maturity after the invention of a new GPT: invention, development, innovation, transfer, growth, competition and consolidation.[119]

Hughes first studied the development of as electrical technologies as they developed through the three phases of electric lighting systems, universal lighting and power systems, and large regional power systems after 1890. Each stage of development was characterized by specific critical problems that attract the attention of characteristic types of problem solvers: inventor-entrepreneurs, manager-entrepreneurs, and engineer-entrepreneur.[120] He applied this life-cycle model to the growth of industries like electricity generation and automobiles, industrial complexes that arose in the first half of the twentieth century from major nineteenth century inventions like electricity and the internal combustion engine.

The characteristics of different stages in the growth to maturity of these technological systems were, for Hughes, the key to understanding how a GPT affects an industry's development. In the industry life cycle, each of the stages a technological system evolves through has a dominant business model and type of person. The business model for an established, mature system is management based, focused on managing the growth of the firm and incremental improvement of their product or service.

Within the seven stages of the industry life cycle are two interior cycles that divide industry evolution into two stages. Cycle 1 is invention, development, innovation, and transfer, Cycle 2 is growth, competition, and consolidation. Invention and development clearly applies to emerging industries going through rapid technological evolution, driven by new GPTs that inaugurate new production systems and development of supporting networks. In Cycle 1 these technologies lead to the creation of new industries, composed of clusters of specialised firms built around the different components of the new GPT. For example, the development of the internal combustion engine led to firms supplying different parts of cars and trucks like chassis, engines, tires, dashboards, and chips that make up the automobile industry, and the electricity industry required firms making generators, dynamos, transmission networks and appliances.

INCREMENTAL INNOVATION IN CONSTRUCTION

The process where inventions are developed, tested and extended, and finally put into production is one of innovation. Firms refine specific parts of a production system, usually in response to something changing elsewhere in the system as production and distribution methods evolve over time, step by step. Although this form of innovation is incremental, it should not be dismissed as unimportant. An example is the steady increase in lifting capacity of tower cranes since their invention in 1949. Another example is the development of computer-aided design (CAD) software, which went on for two decades before Autodesk was started in 1982, one year after the first IBM PC. Over the decades Building information models (BIM) have advanced through 2D and 3D versions to the 4D (schedule) and 5D (cost) iterations today. Now software linked to cameras on helmets or drones can provide real time augmented reality (AR) images from a building site linked to the BIM model of the project.

In construction, many technical advances have come from materials suppliers or component, plant and equipment manufacturers, who have been responsible for the introduction of new products and equipment, such as excavators, cranes, facades and lifts. These are incremental innovations directed at improving existing processes done in construction. Across the construction supply chain firms that develop or exploit new technologies such as lifts and elevators, glass facades, and interior wall systems, don't create new industrial networks. These firms become part of an existing network, which is the construction production system. As a mature system, many of its sub-markets can be expected to be concentrated and oligopolistic, with a few large, well-established firms exactly like those Joseph Schumpeter suggested would be most likely to engage in R&D, invention and innovation.

Concrete is an example of how effective incremental innovation in construction materials can be. By the 1880s the increasingly widespread use of concrete had changed its status from hobby to a modern industry, as scientific investigation into its material properties revealed its shear and compressive characteristics. With the development of reinforced concrete there was change in architectural concepts of structures and approaches to building with concrete. The industrial standards of concrete technology influenced ways of thinking based on building systems and standardized building elements. These became identified with what was known as the Hennebique System, a simple to use system of building with reinforced concrete columns and beams patented in 1892. By 1905 Hennebique's system had spread across Europe and elsewhere and his company employed 380 people in 50 offices with 10,000 workers onsite.[121]

Concrete then set the agenda for the development of construction as a technological system over the next hundred years, driven by the modernist movement in architecture as it explored the possibilities of these material for increasing the height and scale of buildings and modern construction materials and methods spread around the world.[122] For over one hundred years, since Hennebique, there has been ongoing refinement and development of the world's most widely used construction material, as shown in Table 1.[123]

Concrete shows how incremental innovation in materials played a significant role in the reorganization of site production methods as mixers, pumps and chemicals were refined and developed in a long process of interconnected innovations. One of the characteristics of a successful technology are these spillover effects, with advances in one industry leading to complimentary developments in related industries. Current development of 3D concrete printing combines BIM models, new concrete mixtures and chemicals, and new printing machines. Again, a combination of new materials and new machinery is required for this technology to work.

Table 1. Incremental innovation in concrete since 1800

Date	Innovation
1800	Portland cement
1867	Monier reinforcement system
1892	Hennnebique System
1911	Concrete mixer
1913	Ready mix concrete
1918	Duff Abram's water/cement ratio law
1925	Concrete pumps and vibrators
1928	Freyssint prestressed concrete
1938	Air entrained agents
1952	Silica fume
1955	Chemical admixtures, slipforming
1960	Fibre reinforced concrete
1964	Superplastizers and high performance concrete
1980	Roller compacted concrete
1991	Fibre reinforced polymers
2001	Porous concrete
2003	Self-compacting concrete
2011	New generation of sealants and additives

Source: Jahren, P. 2011. *Concrete: History and Accounts*, Trondheim: Tapir Academic Press.

Industry Consolidation and Inertia

The life cycle of a production system starts with the appearance of a new GPT that requires new forms of organization. This reorganization will be slowed by the organizations' business models and people in networks of influence that might be threatened by a new advanced technology.

A mature technological system is a fully developed system of production developed over decades, as experimental work becomes a standard technology.

The built environment and current building and construction products and processes are the outcome of a long development path. Many of the industry's global leaders are well-established, Bechtel for example is over 100 years old, and other firms like Hochtief, Skanska, and AECOM can trace their origin stories back over a similar period. Shimizu is over 200 years old. It's a remarkable fact that construction today is a technological system that has been developing for more than 150 years, since the arrival of steam, steel and concrete. With the early growth stage based on invention of new technology long over, modern construction has since been in Hughes' Cycle 2 late stages of competition and consolidation, where successful firms survive and thrive, and gain both market share and market power.

In an industry life cycle, once past the initial growth stage technology stabilises around standardised products and processes, and the shape of the industrial structure emerges. In many cases such mature industries are oligopolistic, with a few specialized firms dominating market niches or layers in the supply chain. Consolidation leads to industry concentration with large firms dominating their markets. The car industry is an example, where two-thirds of global production is done by eight firms and there are often only two or three key suppliers of dashboards, door panels, seats, airbags, brakes and steering and other key components. Construction materials like cement, concrete and glass, and components like building management systems, interior walls, plumbing fixtures, lifts and elevators are all oligopolistic industries in a mature supply chain.[124]

These later development stages produce a specific culture of technology with distinctive values, ideas, and institutions that organise the necessary knowledge and practice across the industries involved. The technology, its systematized knowledge and its culture become embodied in the firms and social institutions in a mature technical system of production.[125] This culture of technology combines with large-scale organizations and institutions and the careers of practitioners to create technological momentum, the tendency of technologies to develop along defined trajectories unless or until deflected by a powerful external force. A technical system can be characterised by its values and culture, an example is drainage and sewerage systems,[126] because these systems are designed and operated according to accepted rules and practices, connected by standards and professional associations. Architectural historian Howard Davis defined the 'culture of building' as 'the coordinated system of knowledge, rules, procedures and habits that surrounds the building process.'[127]

Economist Brian Arthur examined the possibility that history matters, arguing historical 'small events' are not averaged away and 'forgotten' by the dynamics of industry development, but may decide the eventual outcome. He developed the idea of technological lock-in and persistence, the well-documented persistence of older technologies despite newer and better versions being available, like the QWERTY keyboard or AM radio.[128] Mature systems have inertia from the organizations and people committed by various interests to the system, as Thomas Hughes argued: 'A grievous flaw in the reasoning of enthusiasts for radically new technology ... lies in the former's failure to take into account how deeply organizations, principles, attitudes, and intentions, as well as technical components, are embedded within technological systems.'[129] At the beginning of the twentieth century 'the older assembling industries like engineering were slow to change. Each firm took a proprietary pride in its own work' and the trades 'fearful of technological unemployment, fought all changes in conditions of work.'[130] The Bell Labs 1950s transition from an analogue to digital phone system was described as 'resolved not by persuasion but by attrition, the managers who were analogue advocates died out, and a new generation of digital managers took over.'[131]

Construction of the built environment has momentum and inertia affecting technological innovation. It is a mature, project-based system of production with complex professional, organizational, contractual and working relationships, and is geographically distributed, causing significant horizontal and vertical differentiation within construction firms with potential for uncoupling between project activities and organizational strategies. Innovation is difficult, but not absent.[132] Moreover, the context is one of wider networks containing many small and medium size firms with a range of organizational and institutional relationships, where external contracting is common. All these factors are seen as inhibiting, although not preventing, innovation and diffusion of new technology.

TECHNOLOGY, DEVELOPMENT AND DIFFUSION

How firms utilise technological capabilities differentiates them within a diverse, location-based production system. It is widely recognised there are differences between industries in the way that technology is adopted, adapted and applied, but differences within industries generally get less attention. Technology adoption within industries depends on the different individual characteristics of firms. These affect the rate and extent of adoption of new technologies and the effects of competitive dynamics, which is how the adoption of new technology by one firm influences the adoption of technology by other firms in that industry.[133]

For construction this is significant, not only because of the number of small and medium size firms, but because of the size and reach of the major firms. A global contractor might have over 50,000 employees, suppliers of basic materials and sophisticated components are large multinational or multilocational industrial firms, there are some large global consultants and project managers, many of these firms are publicly listed, and so on. These firms have the management and financial resources required to invest in twenty-first century technology, if and when they decide to do so. The issue here may be the ability of incumbent firms to capture knowledge externalities, adopt new technologies, and adapt to the impacts of emerging technologies and their requirements.

How and why a new technology spreads through the economy and society is determined by many factors, however studies of historical cases such as tractors, electricity, TV and phones have given good examples of technology diffusion and its dynamics. A GPT takes time to diffuse through the economy because parallel changes in forms of organization, methods of production and patterns of consumption are required, and these are not decisions firms and households make quickly or easily.[134] Studies on the introduction of new technologies found it takes 15 to 30 years for a new technology to reach 90 percent of its potential market, for example electrification in the US, took 30 years from 1900 because of the fundamental changes industry and households needed to make to take advantage of electrical power.[135]

Another example is the number of tractors on farms, as the tractor displaced horses and mules in US agriculture between 1910 and 1960. Horses and mules declined from about 26 million in 1920 to about 3 million by 1960, while the number of tractors rose from zero to 4.5 million by 1960.[136]

One reason for the slow spread of tractors was the incremental innovation needed to increase their quality and reliability. A second was labour shortages and an increase in farm wages after 1940. The relative prices of labour and mechanization have been found to be the most significant factor in technological innovation, diffusion and automation.[137]

Technological diffusion in an established system like construction of the built environment is slowed by an interlocking set of factors, the economic, political, legal, and social barriers that make innovating so difficult.[138] Other sectors of the economy have similar entrenched systems of production that resist innovations that might threaten to disrupt their business models. As long as current technology meets the expectations of clients and users for prices and dominant products, there are significant market imperfections such as network economies, lumpiness, split incentives, requirements for collective action, and transaction costs, that inhibit diffusion and can block entry of more efficient, advanced technologies. There is also an institutional structure that imposes regulatory hurdles or other policy disadvantages, favours existing technology or discourages new entrants, with politically powerful vested interests who resist the introduction of threatening technologies supported by a financing system based around incumbents. Educational curricula, career paths, and professional standards use existing technology.[139] For buildings and structures safety is a major issue.

Thomas Hughes was particularly interested in a small group of people he called 'system builders', men like Henry Ford and Thomas Edison, who conceived and built entirely new and fully integrated supply chains around the GPTs of the internal combustion engine and electricity. Those GPTs initiated the production systems of many networked suppliers used to produce cars and electricity, and in Hughes' book *American Genesis* these entrepreneur-inventor system builders have a central role. A key difference is that a technological system like electricity has many suppliers brought together in a few places, and a distributed system like construction has many suppliers in many places. This is one reason new construction technology finds it hard to achieve scale, although a production system as decentralised and diverse as construction has many potential entry points for new technology.

The driver of technological development in construction of the built environment is unlikely to be system builders like Thomas Edison and electricity generation, although there have always been significant individuals in construction, such as I. K. Brunel, Hennebique and Ove Arup, and demonstration projects like the Crystal Palace, Eiffel Tower and Brooklyn Bridge have played an important role because imitation is a powerful force in the spread of innovation. However, over one hundred years later, construction has become a mature production system based around standards and professional roles, with a high degree of technological lock in due to its age. The 'embeddedness' of the system is found across the professional institutes and organizations, trade and industry associations, government regulations and licensing, standards and codes, insurance and finance providers and regulators. Within such a complex system, innovation will typically be incremental and not obvious to the occupants of buildings, and the adoption rates of different technologies will vary greatly.

Hidden Innovation and Technology Adoption in Construction

Measuring innovation in construction of the built environment is not easy. New products are typically based on years of research and development that turn an invention into a usable technology. After the launch of a new product the technology will follow an adoption path, and successful technologies will gain market share over periods ranging from a few years to several decades as products based on them improve and become more widely used. Firms' investment in equipment, software, training and design supports the innovation required to remain competitive, although very little of that investment is included in industry R&D statistics that only include expenditure by research labs and institutions.

National statistical agencies do not provide data on the number of new products or services introduced by firms, most of which are incremental improvements, and do not capture organizational innovations like lean construction or production improvements that increase efficiency, such as the lifting capacity of tower cranes. However, they do provide survey data on household use of new technologies. Adoption paths for a range of technologies by US households are shown in the following figures.

There are two distinctive characteristics of construction innovation and technology adoption that are important. The first is the fragmented, dispersed market for construction, where new products have to gain market share project by project against existing products. The second is the extent of 'hidden' innovation that improves the performance of buildings. Hidden innovation improves performance of a product but is not visible to the user or customer. Examples of hidden new technologies in automobiles are disk brakes and electronic ignition, neither of which would register with most car drivers.

A great deal of construction innovation is 'hidden', either inside the structure or in process innovations in onsite work, such as powered hand tools or new lifting and hoisting equipment. Higher standards for building materials' fire, thermal or seismic performance are not visible, improvements in their air quality or energy use are not obvious to users, and products like ducting, composite wiring and plastic pipes are inside the structure. Hidden innovation is often an incremental upgrade of a current product or system. Process innovations that improve site efficiency do not benefit clients and customers directly, they will generally be captured by the contractor through higher profits (or lower losses) on a project rather than passed on to clients through lower price or better quality.

Piping and plumbing provide good examples of hidden innovation in construction of the built environment. Flushing toilets are one of the most successful inventions of all time and are themselves visible. However, the piping, drainage and sewerage systems that were required, and the pumps and equipment needed to make the system work, are largely out of sight. The use of PEX pipes in plumbing provides another example. PEX pipes are made of cross-linked polyethylene and, as an alternative to PVC or copper, are used for water distribution, heating and cooling. Invented in Germany in 1968, PEX became available in Europe in the early 1970s and in North America in the mid-1980s, where it took 30 years to take half the piping market in residential construction but made little inroads in use for commercial buildings.

Figure 7. Adoption rates for information and communication technologies

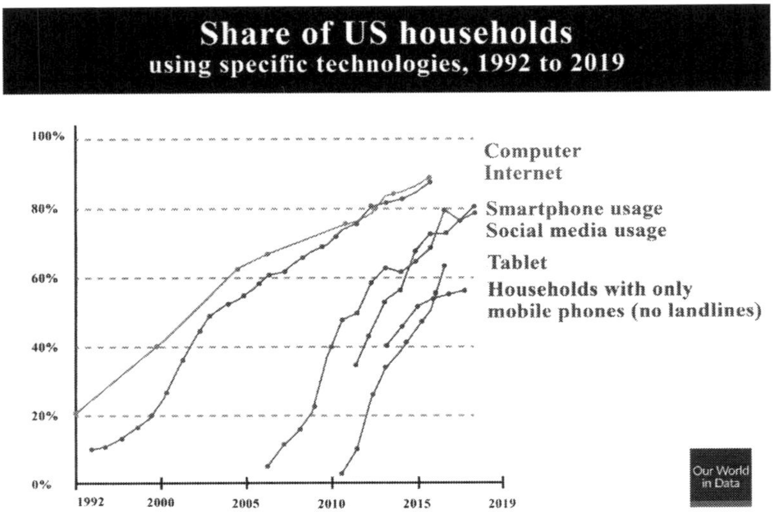

Figure 8. Adoption rates for household technologies

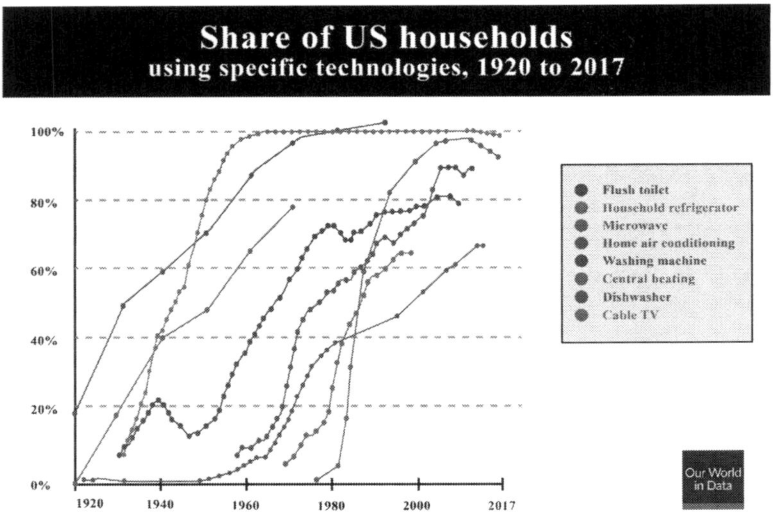

That it takes two or three decades for a new technology to become widely used is generally true, but the adoption path is different for each technology. For example, the speed of uptake of IT by US households was much faster for smart phones and social media than for computers and the internet, with the former reaching 80 percent in a decade while it took 20 years for the latter (Figure 7). Tablets took 5 years to get to 60 percent.

For household appliances there is a similar story, with microwave oven and refrigerator use rising quickly over a decade to nearly 100 percent, while it took 60 years for washing machines and 40 years for air conditioning to reach 80 percent. Dishwashers took 40 years to get to 60 percent of households (Figure 8). All of these would be regarded as successful innovations, but their adoption rates and diffusion pathways varied greatly.

Some of these household technologies required new structural elements in buildings, and to an extent can also be regarded as construction innovations. Refrigerators require electricity, and the first building codes in the 1890s were for electrical systems and components. Flushing toilets, central heating and ducted air conditioning were new functions that became incorporated into the product, and by the mid-twentieth century the basic parameters for building design and materials were in place based on codes that are regularly revised to set higher performance standards. The functionality of housing has not fundamentally changed since, and continuing production, product and process innovations have been hidden from consumers and users.

Oligopolistic industries with a few dominant firms like automobile manufacturing can scale innovations relatively quickly and successful innovations can reach 100 percent usage in a decade or so, as Figure 9 shows. If one manufacturer improves their product with a new technology (the innovator) the others will follow (the imitators), otherwise they risk losing market share. This behaviour gives the S-shaped technology diffusion curve in these figures, rising quickly as imitators take up the technology and flattening out when maximum market share is reached, a dynamic also at work in consumer markets. In less concentrated industries with more manufacturers and dispersed, competitive markets, few technologies will be able to reach 100 percent of their potential market. Plastic water pipes have not replaced copper piping for example. In a few countries automated cabinetry and prefabricated houses have large market shares, but not in others.

Consider the market for external cladding for buildings. To the traditional materials of wood and brick have been added glass, concrete, metal, ceramic, vinyl, composite and fibre cement alternatives, all of which have continuing significant shares of the total market. There are over a dozen major manufacturers who all have large market shares in some, but not all, regional markets (like Europe, East Asia, Central, South and North America). The most recent innovation is HardiePlank, a fibre cement panel invented in Australia in 1986 that took 20 years in the US to reach 20 percent of the housebuilding cladding market.[140]

Figure 9. Adoption rates for automobile technologies

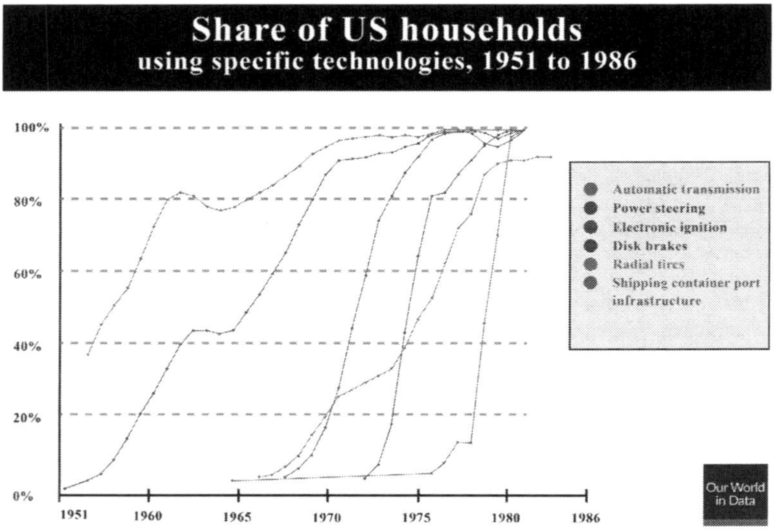

The technology adoption rate in construction might be more like the one seen for household appliances than automobile manufacturing, where the basic function of the product is transport and the industry is concentrated in half a dozen global manufacturers. Construction is a dispersed market with many customers requiring different functions spread out over different locations. Contractors are typically local firms, and their suppliers range from small workshops to giant multinational corporations, allowing opportunities for substitution between products that are inside the structure and hidden from users and consumers. Local markets also have characteristic preferences in design and materials, particularly for houses. With design decisions and specifications made from project to project and familiar products usually favoured, gaining market share in construction for a new product will be a slow process that requires commitment and resources to be successful. This characteristic also means many innovations have never achieved scale or significant market share.

In the second half of the 1800s steam power and the new materials of glass, steel and reinforced concrete arrived, bringing with them new business models and new entrants. Nearly two hundred years later construction is a mature system, based around standards and professional roles, with a high degree of technological lock in due to the age of the system. The 'embeddedness' of the construction technological system is found inside the professional institutes and organizations, trade and industry associations, government regulations and licensing, standards and codes, insurance and finance providers and regulators.

The seven phases of the life-cycle model explains and illuminates the development of the system of production for the built environment in the twentieth century. It became a dense, highly regulated network of industries, with powerful incumbents in many parts of the supply chain, using standardized materials and components to deliver buildings and structures using well understood processes. Incremental innovation is widespread, but not always obvious when dispersed over many projects in many places. The modern system of constructing the built environment may not be elegant, but it is a flexible, sophisticated and resilient system that coordinates many firms in a widely distributed value chain. Because this is an efficient system, any new technology will have to perform extremely well to have any significant effect.

The three dimensions of industry development during the nineteenth century provide a framework for the relationship between technology and construction of the built environment: the industrialization of production, mechanization of work, and organization of projects. They can also be applied to the twentieth century, when industrialized methods of construction arrived based on offsite manufacture, with the promise of again reorganizing production, work and project management.

4 Industrialized Building and Modern Construction

A building is only as good as its client
Norman Foster[141]

IN THE TRADITIONAL craft system of production a master builder would have carpenters, stone masons, thatchers and other trades working onsite, often living there until the project was finished. This changed with industrialized materials that were produced offsite like bricks, iron and glass, and contractors set up centralized workshops to supply their projects. Since the first industrial revolution the balance between onsite and offsite production for construction has shifted backwards and forwards due to variations in demand, prices, skills and technology.

During the 1830s gold rushes in Australia and California kit homes were exported from the UK, along with a variety of other buildings such as churches and civic buildings. Made of pre-cut timber these were among the first manufactured buildings, meeting demand from increasing populations and solving the problem of local skills shortages. Famous projects like the 1851 Crystal Palace and 1889 Eiffel Tower were made of prefabricated components made offsite.

In the first half of the twentieth century there were many attempts by architects to turn housing into a manufactured product. Le Corbusier thought his Maison Citrohan would revolutionize housing in 1920. Frank Lloyd Wright had his idea of 'Assembled Houses' in 1932. Buckminister Fuller patented the Dymaxion House in 1928 and the Dymaxion Bathroom in 1938. An airplane manufacturer made a prototype which became known as the Wichita House in 1944, but the cost of setting up a production line meant none were produced.[142] Architectural historian Gilbert Herbert's book *The Dream of the Factory-made House*[143] was about a 'Package House' to be mass-produced in a factory making 10,000 houses a year.

Set up in 1942 the factory made fewer than 200 houses, most of which were unsold, and ended up making sets and scenery for Hollywood films. Although architects were influential in design, their ideas were not successfully turned into businesses building houses.

This does not mean there were not many manufactured houses made. Over 100,000 kit houses were built in the US between 1908 and 1940, most of them balloon frame kit homes sold by Sears Roebuck (75,000) and Gordon-Van Tine (20,000), although there were four other large and dozens of small kit home makers.[144] In post-war France the Prouvé Workshop delivered hundreds of 'Meudon Houses', and in the UK more than 150,000 prefabricated homes were built between 1945 and 1948. In the US a developer called William Levitt created Levittown in 1947 in New York. Wanting to be the 'Henry Ford of the building business' his production line had 26 steps and his houses could be assembled in a day. They were cheap and sold by the tens of thousands. Eventually there were five Levitttowns.[145]

Brutalist architecture had emerged in the 1950s in the UK during reconstruction, a minimalist architectural style that used precast concrete and modular elements for multi-story buildings. In the 1960s a group of UK architects known as Archigram first used 'capsule' and 'pod' instead of house or bathroom, and envisaged massive high rises made up of these mass-produced units. By the 1970s the idea that prefabrication was the solution to problems of poor quality and low productivity in construction became central to a movement to 'reform' construction by making it more like manufacturing

The production of building elements and components somewhere other than the construction site has been variously called prefabrication, pre-cast and pre-assembly construction, and offsite manufacturing (OSM). The degree of OSM and preassembly varies from basic sub-assemblies to entire modules. The use of OSM varies greatly from country to country. Types of offsite construction are panelised systems, volumetric systems with partial assembly of rooms, units or pods offsite, and factory built modular components or homes. Offsite manufacture is here used to describe factory production and preassembly of components, elements or modules. Prefabrication is used to describe offsite production of components that are installed onsite.

The chapter next discusses the construction reform movement, then reviews OSM. This is followed by four short case studies of industrialized building:

Sears Modern Homes, Legal & General Homes, Katerra and the Japanese automated building systems of the 1990s are described. The chapter concludes with current developments in software platforms for integration of design and construction.

Productivity and Reforming Construction

The rate of growth of productivity across the OECD first became an issue in the late 1960s, when declining output per hour worked and output per person employed became the focus of a large research programme that sought to interpret and analyse the causes of what became known as the productivity slowdown. The construction industry's low productivity growth attracted attention because its rate of growth of productivity was poor,[146] even by comparison with a long-run overall industry average in the order of two to three per cent a year. Since the 1960s construction productivity has continued to show little or no growth.[147]

In their responses to the lack of growth in construction productivity, governments tried to encourage innovation while keeping costs as low as possible. As the largest client it is not surprising the construction reform movement was led by governments with inquiries, commissioned research and funding for demonstration projects. Many countries developed industrialized building systems for social housing and institutional building projects, and clients began using design and build for procurement more often.[148] Sophisticated procurement systems were developed, such as partnering, alliancing and framework agreements, while major projects got larger and more complex and megaprojects over US$1 billion became more common. Client groups like chambers of commerce, business associations and industry bodies became active, with roundtables, reports and participation in government inquiries. Despite these efforts made by governments, industry organizations and firms over the decades, the rate of growth of construction productivity as measured by national statistical agencies remained low compared to many other industries, particularly manufacturing.

The reasons given for this stagnant growth of productivity are various and include volatility of demand, fragmentation and the number of small firms, the one-off nature of projects,[149] the high labour intensity of residential building, poor economies of scale, limited competition, regulatory impediments, a lack of innovation, poor management, low levels of capital investment and of skills.[150] The rates of technology adoption and diffusion has always been an issue.[151]

As the example reports in Table 2 show (a small sample of the hundreds, probably thousands, of reports done between 1960 and 2000), efforts to reform construction and improve productivity were an international movement that also included Japan,[152] Canada,[153] and Hong Kong.[154] Countries generally followed two strategies, although with considerable variation between them. The first focused on contractual relations between clients and main contractors. The second was industrialized building and offsite manufacturing (OSM). Many systems were used for OSM in different countries, and very many buildings successfully constructed using those systems in the decades after the 1960s, however OSM has not replaced onsite construction but found niches in specific building types and components.

The UK reform movement is particularly well documented, there are a dozen reports summarised and discussed in Murray and Langford's *Construction Reports 1944–98*. They concluded those reports agreed on the poor performance of construction, had minor differences between their explanations for poor performance, and gave similar recommendations for improvement Although the reports discussed many issues, such as productivity, quality, training, contracting and documentation, the fundamental issue was the cost of construction, reflecting the UK government's role as both a major client and the initiator of the inquiries and research. The last two of those reports by Latham in 1994 and Egan in 1998 became particularly influential as the UK government became a leading advocate of reform.[155]

Table 2. Examples of construction reports from the 1980s and 1990s.

United States
CERF, 1991. *Transferring Research into Practice: Lessons from Japan's Construction Industry*, Civil Engineering Research Foundation, Washington.
Construction Industry Institute, 1986. *An Analysis of the Methods for Measuring Construction Productivity*. Austin.

National Research Council, 1986. *Construction Productivity: Proposed Actions by the Federal Government to Promote Increased Productivity in Construction*, Building Research Board, U.S. Department of Commerce, National Academy Press, Washington.

Construction Industry Institute. 1985., *Attributes of Material Management.* Austin.

Business Roundtable, 1985. *More Construction for the Money*, New York.

Business Roundtable, 1982. *Measuring Productivity in Construction,* Construction Industry Cost Effectiveness Report, New York.

United Kingdom

Egan Report, 1998. *Rethinking Construction*, Department of Environment, Transport and the Regions, UK. HMSO, London.

Latham Report, 1994. *Constructing the Team: Final Report of the Government/Industry Review of Procurement and Contractual Arrangements in the UK Construction Industry*, HMSO, London.

NEDO, 1989. *Promoting Productivity*, National Economic Development Office, London.

CSSC, 1988. *Building Britain 2001,* Centre for Strategic Studies in Construction, Uni. of Reading,

NEDO, 1976. *Engineering Construction Performance*, National Economic Development Office, London.

Singapore

CIDB, 1992. *Raising Singapore's Construction Productivity*, Taskforce Report, Construction Industry Development Board

CIDB, 1989. *Cost Competitiveness of the Construction Industry in Singapore.*

(In 1998 the CIDB became the Building Control Authority)

Australia

Productivity Commission, 1999. *Work Arrangements on Large Capital City Building Projects*. Ausinfo, Canberra.

DISR 1998 *Building for Growth*, Dept. of Industry, Science and Resources, Canberra.

CIDA, 1995. *Measuring Up or Muddling Through: Best Practice in the Australian Non-Residential Construction Industry*, Construction Industry Development Agency.

Industry Commission 1991. *Construction Costs of Major Projects*, AGPS, Canberra.

RCBI, 1991. *Productivity and Performance of General Building Projects*, Royal Commission into Productivity in the Building Industry in NSW.

DITAC, 1990, *Survey of Project Performance*, Dept. of Industry, Technology and Commerce, Canberra.

DITAC, 1989, *Indicative Survey of Non-Residential Construction Industry Efficiency*, Dept. of Industry, Technology and Commerce, Canberra.

Egan began his report arguing industry improvement required changing the industry culture, recommending lean production methods using examples from car manufacturing, steel-making, grocery retailing and offshore engineering, and setting ambitious performance targets for the industry.[156] The government followed Egan by promoting offsite manufacturing in the *Modernising Construction* and *Accelerating Change* reports,[157] and supported the reform movement with legislation and by funding three agencies. These were Rethinking Construction, Construction Best Practice and the Movement for Innovation, which were brought together in 2004 as Constructing Excellence with the mission 'to achieve a step change in construction productivity by tackling the market failures in the sector and selling the business case for continuous improvement.' Through programmes in innovation, best practice knowledge, productivity and engagement, Constructing Excellence developed a 'strategy to deliver the process, product and cultural changes that are needed to drive major productivity improvements in the sector'.

In the UK government promotion of industrialized building became associated with the construction reform movement, which had traditionally focused more on contractual issues. Following Egan comparisons of productivity growth in construction and manufacturing were made, targets were set, and numerous agencies and bodies formed to push reform. The history of efforts to reform the UK construction industry is documented in numerous reports, many commissioned by the government of the day. Over the decades that history shows how ineffective these efforts and policies were in changing industry culture and practice, with a 2016 report called *Modernise or Die*[158] again arguing for OSM as the solution.

Issues with Offsite Manufacturing

Many buildings have been, and are, built using OSM. Hotels and fast-food chains use OSM for their buildings. Governments have promoted it since the 1960s, and there have been large scale public housing projects using OSM in many countries. However, despite successes like Singapore's Housing Development Board, and European and Japanese manufactured housing, OSM and industrialized building generally have a small share of the market for total construction.[159]

Advocates claim the benefits of OSM are increased control of construction processes within a controlled environment for handling and storage of materials. While reducing waste and improving quality, there will also be reductions in unit cost as the scale of production increases.[160] To support this positive view the development of the automobile industry with lean production and the Toyota method was often used, and Japanese prefabricated housing builders provided as an example of successful OSM.[161] A review of research in the UK and United States identified the drivers and constraints of OSM, and the tables below summarise the findings and provide good overview of the general position of OSM in the 2000s.

Table 3. Drivers of offsite manufacturing identified

Driver	Description
Process & Programme	Less time onsite, speed of construction and delivery of product increases Less time spent on commissioning Guaranteed delivery, more certainty, reduced management time Programme driven centrally. Simplifies construction process.
Quality	Higher quality from factory. Product tried and tested in factory Greater consistency—more reproducible More control of quality, consistent standards, reduced snagging and defects
Cost/Value/ Productivity	Lower preliminary costs. Increased certainty, less risk Increases added value. Lower overheads, less on-site damage, less wastage Reduced whole-life cost
People & OHS	Fewer people on-site – possibly reducing OHS risks
Skills & Knowledge	People know how to use products Limited or very expensive available skilled on-site labour

Logistics & Site Operations	More success at interfaces. Less site disruption Reducing the use of wet trades. Removing difficult operations off-site Work continues on-site independent of off-site production and vice versa Restricted site layout or space limits site operations. Site restrictions by external parties alleviated
Environmental Sustainability	Reducing environmental impact during construction Maximising environmental performance throughout the

Source: CRC. *Offsite manufacture in Australia: Final report.* 2007: 18.

After decades of efforts to promote OSM, the constraints have outweighed the drivers and benefits. Note the 'Deep rooted pessimism over past mistakes rather than a determination to learn from history', and 'Client resistance, often due to negative image' above. At this stage the market share of OSM remains small and niche, estimates are low single digits of total construction work in the UK, US and Australia. Success elsewhere is restricted to a few specific markets and project types. The problem was not the technology, which can be made to work, but the expected economies of scale are difficult to achieve because of a range of factors. Some of these factors are internal to construction, like the constraints below, but others are external. In particular, macroeconomic events like financial crises or energy and commodity price changes can quickly undermine the OSM business model.

The US and UK governments supported OSM, with the UK government supported research, publishing case studies and anecdotal evidence promoting OSM in construction for decades.[162] In the US a Technology Roadmap for Advanced panelised construction was produced in 2003 for the Department of Housing and Urban Development as a Partnership for Advanced Technology in Housing (PATH[163]). Despite these efforts, offsite production is not industry practice in either country. Although pre-cast concrete and panelised construction is widely used, OSM has not led to significant advances in mechanization or required a thorough reorganization of project management methods.

Table 4. Constraints to offsite manufacturing identified

Constraint	Description
Process & Programme	Longer lead-times Inability to freeze design & specification early Variations cannot be easily accommodated in OSM Key decisions preclude OSM with poor fit between design and OSM
Cost/Value/ Productivity	Seen as expensive when compared to traditional methods. High initial cost. Obliged to accept lowest cost rather than best value Obliged to accept element-specific costing Clients having difficulties understanding the benefits
Regulatory	Restrictive, fragmented, excessive, onerous, costly between jurisdictions Few codes and standards available
Industry & Market Culture	Deep rooted pessimism over past mistakes rather than determination to learn from history

	Client resistance, often due to negative image.
	Clients view OSM as standardised and lacking customisation
	Strong client perception that OSM, quick-built products are of lower quality
	Resistance by labour (especially unionised) to change
	Resistance to change by builders due to desire for independence
	General construction industry fragmentation
	Difficult to obtain finance from institutions not familiar with OSM
Leadership	Lack of visionaries committed to change in the industry
Supply-chain & Procurement	Unwilling to commit to single point supplier (increased risk)
	Limited choice of supply-chain for the project
	Limited capacity of supplier(s) or supply not available locally
	Inter-manufacturer rivalry preventing the development of a common framework and interchangeability of products.
	Lack of standardisation especially at interfaces
Skills & Knowledge	Professional skills/knowledge:
	-General lack of systems engineering and systems analysis
	- Limited OSM experience, a lack of manufacturing and engineering skills
	- Limited expertise in off-site inspection, poor quality product assurance

	Site skills/knowledge: - Lack of familiarity with OSM systems, lack of trades and training schemes - Lack of IT knowledge and tools by small builders to improve process Offsite skills/knowledge: - Product or component repeatability not feasible due to low volumes - Difficult to re-use process on new projects - Concerns over intellectual property rights
Logistics & Operations	Problem transporting large, heavy manufactured products to site Limitations to movement of OSM units around site

Source: CRC. *Offsite manufacture in Australia: Final report.* 2007: 19-20.

With the development of building information modelling (BIM) and design for manufacture and assembly (DfMA) after 2010, advocates of OSM could argue technology would fix the problem of capital cost and economies of scale. There is little evidence to support that argument. An example is the concrete panels for Crossrail stations in London. Laing O'Rourke 3D printed 1,400 unique moulds to make precast glass-fibre reinforced concrete panels for three Crossrail stations. These were used to cast 36,000 different shapes and printed by a 6-axis gantry robot in a space 30 metres long by 3.5 metres wide by 1.5 deep. The robot then polished the moulds' surface with milling tools, combining additive and subtractive methods.[164] While a successful demonstration of the technology, it was not a successful contract for Laing O'Rourke.

Four Cases of Industrialized Building

There are many examples of successful prefabricated projects and companies that manufactured thousands of homes. However, the four examples discussed below of Sears Modern Homes, Legal & General Homes, Katerra and the Japanese automated building systems of the 1990s illustrate the problematic nature of OSM. The first three manufactured housing, and the Japanese systems made commercial office blocks. An interesting characteristic of industrialized building has been the entry by large firms, sometimes from outside the industry, who had the capital to invest and an appetite for risk. In 1908 it was retailer Sears and Roebuck, and in 2019 Amazon invested in a Californian housing manufacturer.

SEARS MODERN HOMES

A century before Ikea sold their first prefabricated Boklok house in 1996, one fifth of Americans were subscribers to the Sears and Roebuck Mail Order Catalogue. Anyone anywhere in the country could order a copy for free, look through it, order something, and have it delivered to their door by the newly built railway network. At its peak the Sears catalogue offered over 100,000 items on 1,400 pages, and in 1908 they began offering houses. Although it was not the first company to sell kit homes by mail order, Sears came to dominate the mail-order market. Between 1908 and 1940 it delivered 75,000 homes.[165]

Sears sold complete houses, called 'kit homes'. Customers selected from one of dozens of different models, they could order, send a check, and a few weeks later a train would arrive with a door secured by a small red wax seal. The new owner would open the boxcar to find everything needed to build the home (including more than 10,000 pieces of framing lumber, and 20,000 cedar shingles for the roof). The lumber came pre-cut with an instruction booklet, and Sears promised that, without a carpenter, a person could finish their mail order home in 90 days.

Then, in 1911, Sears began offering mortgages to their customers. The Sears home mortgage program became one of the keys to their success, as all those homes and their new, mostly young homeowners, needed furnishing and decorating. By lowering the barrier to entry, it allowed Sears to sell more kit homes than any of its competitors, but when the Great Depression came in 1927 many of the mortgages defaulted, and in the

1930s the company ended up foreclosing on tens of thousands of its customers. It was a financial and public relations disaster. After years of declining sales, Sears finally closed its Modern Homes department in 1940.

A few kit home manufacturers that hadn't sold mortgages survived, but Sears was gone. Over the following decades the next housing boom saw the rise of the suburbs and the prefab home As demand surged in the 1950s, factory built homes went up on subdivisions by the the thousands. Companies had varying degrees of success, such as Lustron Homes (made 2,500 enameled steel houses between 1947 and 1950, closed due to increased steel prices), and National Homes Corporation (started 1940, closed 1972 because of rising oil prices and recession).

The case is not that industrialized building doesn't work.[166] Both Sears and National Homes factory built around 2,000 houses a year for over 30 years, supplying new suburbs on the subdivisions around cities. They delivered on the quality and affordability promised by OSM, and many of these houses are still standing. National Homes delivered modules to builders who prepared the site, assembled the house and completed the fit-out, a model successfully used today by Ikea in Northern Europe for their Boklok range of houses. Japan has produced prefabricated and manufactured homes for many years and the industry there is well established. However, industrialized building requires large capital investments and has high operating costs. In an industry where demand is closely linked to economic conditions, high cost factories are as much a liability as an asset.

JAPANESE AUTOMATED BUILDING SYSTEMS

Beginning in the late 1980s, the Japanese 'Big Five' contractors developed eight automated building systems, seven designed for structural steel and one for precast concrete construction.[167] These systems were quite possibly the crest of applications of third industrial revolution technologies to construction, creating an enclosed factory environment for high rise building using computers to control materials and components. The systems had many features in common, with an overhead shelter or hat-truss, a jacking system, a material handling system, and a central control station. The overhead shelter or hat-truss was the top floor and was constructed at ground level first, then elevated using a jacking system. Once a floor was complete, the hat-truss was jacked up again for the next floor. Materials were moved horizontally and vertically by a handling system of automated lifts and conveyors. The Obayashi Big Canopy worked by lifting prefabricated material from the ground floor to the constructing floor and conveying it to the installation point by gantry cranes fixed to the underside of the canopy. The manoeuvring of these components was done using a handheld joystick control.

Table 5. Japanese automated building systems

System	Company	Year	Structure
Push-Up	Takenaka Corp.	1989-1991	Structural Steel
SMART System	Shimizu Corp.	1991-1994	Structural Steel
ABCS System	Obayashi Corp.	1991-1994	Structural Steel
T-Up System	Taisei Corp.	1992-1994	Structural Steel
MCCS	Meada Corp.	1992-1994	Structural Steel
Akatsuki 21	Fujita Corp.	1994-1996	Structural Steel
AMURAD	Kajima Corp.	1995-1996	Steel-Reinforced
Big Canopy	Obayashi Corp.	1995-1997	Pre-cast Concrete

Source: Cousineau, L. and Miura, N. 1998: 51. *Construction Robots: The Search for New Building Technology in Japan.*

LEGAL AND GENERAL MODULAR HOMES

Legal and General is a 180 year old British insurance and investment management firm. In 2016 they announced plans to manufacture homes, targeting affordable housing, and set up a business called Legal and General Modular Homes and set up a 550,000 square foot factory with the capacity to produce 3,000 homes per year, employing several hundred people.[168] However, the opening of the factory near Leeds had many delays. Although the first units were delivered mid-2017, regular production was not achieved until 2919. This is the largest OSM factory in the world and a substantial bet on the future of housing in the UK.

The business model combines production with a development pipeline, where 'institutional investors are the long-term holders of the assets working alongside the best-in-class affordable housing operators who will provide the highest-quality housing management'. Some developers with land banks were purchased and L&G have been preparing a pipeline of developments with build-to-rent and build-for-sale homes, including L&G retirement villages with 1,100 homes. By one account, L&G's total investment in the factory, developers and sites by 2019 was £1.5bn. Production levels in 2021 were in the hundreds not thousands of homes and accumulated losses since 2016 were over £100m.[169]

The capital investment required for OSM and manufactured housing has often led firms to become developers in their own right, as creating a pipeline of work to keep their factory busy seems the obvious answer to the problem of scale. Sustained success with this business model has been elusive. However, in 2022 L&G delivered 450 houses to their development sites and had another 450 awaiting planning approval, with annual production of 3,000 houses the target for 2024-25.

KATERRA CONSTRUCTION

Katerra was Californian start-up, founded in 2015. In 2017 it reached a USD$1 billion valuation. The company's goal was complete vertical integration of design and construction, from concept sketches to manufacturing cross laminated timber (CLT) panels and then delivering and assembling the building onsite. On their projects the company wanted to be architect, offsite manufacturer and onsite contractor. This led to issues with the developers and contractors the company was dealing with, most of whom didn't want the complete end-to-end service Katerra offered.

The company started by developing software to manage an extensive supply chain for fixtures and fittings from around the world, but particularly China, and then added a US factory making roof trusses, cabinets, wall panels, and other elements. In 2016 the business model changed because architects weren't specifying Katerra's products. Katerra would design its own buildings and specify its own products. In 2017 it built a CLT factory that increased US output by 50 percent. The factory shut in 2019. Dissatisfied with design software that didn't meet its needs, it developed a custom suite called Apollo.[170] This was to be a platform for project development and delivery, well beyond the document control and communication of then available software from Oracle Aconex, Trimble Connect, Procore and SAP Connect. Apollo integrated six functions:

1. Report: use an address to find site information, zoning, and crime rates;
2. Insight: design with their two building platforms;
3. Direct: a library of components used in the building;
4. Compose: for coordination between the different groups working on a project;
5. Construct: for construction management:
6. Connect: for managing the workforce on a project, with a database of subcontractors.

One of the company's three founders was a property developer, and his projects provided the pipeline of work that initially made the company viable. A second founder had a tech venture capital fund, the third and CEO did a stint at Tesla. Their ambition was to leverage new technologies to transform building by linking design and production through software, designing buildings in Revit and converting the files to a different format for machines in the factory to manufacture.

At first their buildings were designed by outside architects, then in 2016 the company started a design division. In 2018, after raising $865 million led by SoftBank's Vision Fund, Katerra acquired Michael Green Architecture, a leading advocate of CLT, and over a dozen other architects and contractors. In 2020 the business model changed again, with Katerra taking equity stakes in developments to boost demand. Over five years Katerra had gone through four different business models as they sought to achieve sufficient scale to keep their factories busy.

There were accumulating losses completing these projects due to delays in and large cost overruns during the Covid pandemic in 2020-21. Despite the funding they had received these losses eventually overwhelmed the company, and in June 2021 Katerra Construction filed for Chapter

11 bankruptcy. The failure of Katerra was a reality check for advocates of modern construction methods. It had no lack of capital, but did not solve the demand-side problem of cyclic fluctuations or the supply-side problem of total delivery costs.

In six years Katerra had grown to a 7,500-person company. That growth cost both money and focus, of the total US$2.2bn raised SoftBank had invested $2bn between 2018 and 2020. Without a focus, Katerra didn't clearly target a customer base for their residential developments. The company diversified into a range of fixtures and fittings, got distracted by software, and began developing internet-of-things (IoT) technology. The executive team was dominated by industry outsiders, but Katerra hired architects and engineers from traditional firms, so tension was inevitable.[171] The fatal problem was execution, Katerra didn't vertically integrate acquisitions into a company that did everything. It was fragmented and didn't have a product platform or Apollo ready in time.

The demise of Katerra does not mean digitised prefabrication is not continuing to develop. Buildings made of or with engineered wood are common in Northern Europe and no longer unusual elsewhere. In the US Blokable makes modular housing pods designed to be stacked and connected in a factory, and Plant Prefab factory makes a building to the client's design. A number of companies are pursuing the 'kit of parts' approach, using prequalified fabricators to factory produce standardised elements that integrate production and assembly; examples are Juno, Modulous, Project Frog and PT Blink. By 2021 software rather than factories was becoming the focus of construction technology start-ups and initial public offerings (IPOs).

Platforms, Procurement and Production

With Apollo, Katerra was actually behind a number of other companies developing platforms that manage design and construction in various ways. These platforms are at the technological frontier, a fourth industrial revolution technology for OSM with automated production of components. Platforms are IT systems that provide access to the different software packages used in construction for BIM, document exchange, transactions, project management and monitoring, they aggregate data and process control in a single place.[172] Autodesk, Trimble, Bentley, OracleAconex and Procore are examples, but many contractors, consultants and other large firms in the value chain in materials, manufacturing and distribution have developed internal platforms to manage their operations.

A closed platform is internal to a firm, an open one allows access along the value chain. The linking of online design to local production is known as cloud manufacturing,[173] and it may be 'the threat is that architects and engineers could lose agency in project development to platforms developed by tech companies.'[174] On the other hand, they would be applying their expertise to evaluate the hundreds of solutions produced by generative design systems.

In 2018 Californian firm Project Frog launched their KitConnect platform, bringing together a decade of development into prefabrication and component design, integrating BIM with DfMA and logistics. Project Frog developed a market intermediary platform that combined with Autodesk's design software. The designs are linked to a semi-open marketplace where approved merchants can price and manufacture the products required for a project. Platform design relies on a library of components or 'digital kit-of-parts' for a building. Such platforms are in the process of becoming a basic part of construction tech.

In a 2019 study of three US construction firms, each firm had a different approach to digital manufacturing: DPR with a relational, project-based spinoff; RAD with vertical integration; and Project Frog with their digital system. These firms were integrating OSM and automated production through development of digital platforms that provide design, component specification and manufacturing, delivery and on-site assembly. Hall, Whyte and Lessing concluded:

> ... future platform development will tend to be open or closed, depending on the level of vertical integration for the firm. Open platforms will be developed by digital systems integrators such as Project Frog. These firms will develop the platform core and leverage the principles of industry 4.0 to organize the periphery into new digital ecosystems. Closed, internal platforms will be developed by vertically integrated firms such as RAD. These firms gain advantage from total control of system architecture and the ability to push the limits of technical change.[175]

A platform not only provides connectivity and possibilities for exchange, it gives rise to a new data network by putting a 'computer in the middle of every transaction'.[176] AI-enabled platforms become flexible infrastructures that are capable of learning about interconnections between products, people, and organizations using data collected on their transactions. AI processes can then improve the functionality of the platform, learning from data to improve prediction. Prediction is the ability of a system to draw on existing data from the past to generate information about the future.[177] The greater the speed and accuracy of predictions, the higher their value is to users, so increasing the quantity and quality of data used for training of machine learning algorithms on the platform improves predictions and, therefore, increases the value of the platform. The utility of a platform is driven both by scale and by improvements from data driven learning.[178]

An alternative to the industrialized, vertically integrated contractor/developer model that L&G, Katerra and others have used is software platforms linking design and design libraries with production by prequalified fabricators.[179] Platforms are a fourth industrial revolution technology for OSM that enables automated production of building elements and components. These digital platforms provide designs and specifications for manufacturing, often through a standardized kit-of-parts, and they can play crucial role in the development of AI for construction as the volume, variety, and veracity of data increases, allowing predictive models to improve and data driven AI learning to develop powerful network externalities.[180] Platform based industrialized building is inherently more flexible than a factory based, capital intensive system of production. By aggregating buyers of building components a platform solves the fundamental problem of OSM, which is maintaining a sufficient volume of production at a factory's break-even level or more.

Modern Methods of Construction

Offsite manufacturing, modular and prefabricated building have been transforming construction like nuclear fusion has been transforming energy. These 'modern methods of construction' have a dismal track record. The brutal economies of scale and scope in a project-based, geographically dispersed industry subject to extreme swings in demand have bought periods of their growth and development to an end. There have been successful projects and, in a few countries, manufactured housing has a large market share, but often macroeconomic factors have undermined their viability in the long run.

The idea of construction as production[181] was based on OSM, but after decades of development OSM has yet to become a viable business model beyond niche markets and specific projects. OSM markets exist mainly in housing and institutional building, wherever it is the most effective or efficient piece of technology available and there is a lot of repetition from project to project. This manufacturing-centric view of progress in construction, endorsed by numerous government and industry reports, was the end point of the development trajectory from the first to the third industrial revolutions.

The technological base of OSM is a mix of those from the first industrial revolution, like concrete, with second and third revolution technologies like factories and lean production. Despite all efforts this has not become the primary system of construction of the built environment because OSM does not deliver a decisive advantage over onsite production for the great majority of projects. Instead, construction has a deep, diverse and specialised value chain that resists integration because it is flexible and adapted to economic variability.

The up-front capital requirements of OSM make it a capital-intensive form of production, which brings high fixed costs in an industry characterised by demand volatility over the business cycle. This means macroeconomic events often determine the success or failure of the underpinning business model and the success or the eventual failure of the investment. A batch of new US prefab housing firms failed during the recession after the financial crisis in 2007, demonstrating the importance of the relationship between economic and business conditions and the tenuous viability of the business model for industrialized building.

Manufactured housing in the US also provides an insight into the institutional barriers to industrialization in construction that exist in many countries and cities. Although the Department of Housing and Urban Development has a national code, US cities discriminate against manufactured housing through local and county government planning codes that restrict or ban their use, and often places them far from amenities such as schools, transport, doctors and jobs. Despite these barriers, in 2021 there were 33 firms with 136 factories that produced nearly 106,000 manufactured homes in the US.[182]

While the history of prefabrication features major projects like the Crystal Palace and more recently the Oresund Bridge in 2000, the reality is that OSM has only been successful in specific niche markets such as fast food outlets, hotels and institutional buildings with a lot of repetitive elements, or house manufacturers like the Japanese and Scandinavian firms Sekisui and Ikea. Failures like Katerra in 2021 and Sears Roebuck eighty years earlier have been more typical. Included in the UK 2017 Industrial Strategy was Construction, with the aim to 'change the way buildings are created' with a manufacturing hub for offsite and modular construction. By 2021 the UK Government had moved on, to a kit-of-parts approach focused on the energy efficiency of buildings and new design standards. Similar initiatives elsewhere have also lost momentum and had little long-term effects on the use of OSM and the wider construction industry.

The development of industrialized building and modern methods of construction in the twentieth century did not displace the contractor-led model of onsite production. This may, however, not be true of the twenty-first century if procurement were to shift to online platforms that integrate design, manufacturing and construction based on a digital model of the project.

5 Construction 4.0, AI and Digital Fabrication

Any sufficiently advanced technology is indistinguishable from magic
Arthur C. Clarke[183]

THE IDEA OF a fourth industrial revolution led naturally to discussion of 'Industry 4.0'. The first country to have an Industry 4.0 plan was Germany in 2011, followed by any country with a manufacturing industry, as innovation, science and technology policies directed at these technologies were launched.[184] The World Economic Forum describes Industry 4.0 as technologies that fuse 'physical, digital and biological worlds' in 'cyber-physical systems' that will be a 'new chapter in human development', and gives space technologies, blockchain, IoT, AI and robotics as examples.[185] To that list can be added other frontier technologies like quantum computing, genome editing and nanotechnology. What is unusual about the present is the large number of technologies advancing rapidly; in the past there would only be a few new technologies developing at any one time but now there is a broad front of simultaneous advances that mutually reinforce each other and underpin the fourth industrial revolution.

Around 2018 the amount of venture capital going into construction tech startups began to significantly increase in the EU and US. A 2021 research report by Kabri Construction Research found 300 construction startups since 2000 had raised over US$13 billion, of which $2 billion was raised to 2017, $6 billion between 2018 and 2020 and $5.5 billion in 2021, with nearly $9 billion expected in 2022. The top five companies funded were View (smart glass), Katerra (builder), EquipmentShare and Workrise (platforms), and Procore (CM software). These five got nearly 40 percent of total funding, and the top 20 companies received more than 60 percent of the total. A 'substantial number of these top companies are either high-profile failures (Katerra, Blu, Airware) or are in fairly rough shape (View, Tophat).'[186] A nice example of creative destruction happening on the technology frontier.

Table 6. Top 50 US and European construction startups to 2022

Company	Category	Funding Raised $ million
View	Building materials	1732
Katerra	Builder/developer	1645
EquipmentShare	Marketplace	1554
Workrise	Marketplace	694
Procore	Cnst management software	601
Veev	Builder/developer	585
Icon	3D printing	436
Halio	Building materials	318
Prescient	Builder/developer	295
Built Technologies	Fintech	289
Infra Market	Distribution	270
Blu Homes	Builder/developer	198
Readaptive	Building energy use	197

Openspace.AI	Data capture/ digital twins	190
Tul	Distribution	185
Path Robotics	Robotics	171
RenoRun	Distribution	163
HomeBound	Builder/developer	128
Hover	Data capture/ digital twins	122
Airware	Data capture/ digital twins	120
Matterport	Data capture/ digital twins	114
Span.io	Building energy use	112
Versatile	Data capture/ digital twins	109
Bloc Power	Building energy use	105
Block Renovation	Maintenance & building mngt	104
PlanRadar	Cnst management software	103

Mighty Buildings	3D printing	102
Curbio	Maintenance & building mngt	101
Tophat.io	Builder/developer	98
Biomason	Building materials	93
Billd	Fintech	90
Nexii	Building materials	85
Jobber	Marketplace	84
Super	Other	80
Factory_OS	Builder/developer	78
Boston Metal	Building materials	76
Cover	Small buildings/ office pods	73
PlanGrid	Cnst management software	69
Mosaic	Marketplace	69
Schuttflix	Marketplace	66
Dandelion Energy	Building energy use	65
FinalCAD	Cnst management software	62

Orbital Systems	Other	61
Intellihot	Building energy use	61
Kebony	Building materials	59
Dronebase	Data capture/ digital twins	59
Foresight	Other	58
Diamond Age	3D printing	58
Doxel.ai	Data capture/ digital twins	57
ConXTech	Building materials	56

Source: Kabri Construction Research.[187] Includes only publicly announced funding.

This funding is significant because investment in construction technology startups is a proxy for the development of IP and other forms of intangible capital.[188] In 2021, the IPO for Procore raised US$635 million at a valuation near $10 billion, a record for construction tech. Australian rival Aconex had been bought by Oracle in 2017 for US$1.2 billion.[189] US start-ups that followed in the wake of Katerra like Juno and Generate didn't build factories but outsourced production to prequalified suppliers. Another called Outfit offered homeowners a DIY renovation from its website, ordering and shipping the materials and providing instructions for completing the work (the Sears model again). In 2022 funding was going to 3D printing (e.g. Icon, Mighty Buildings), data capture/digital twins (e.g. Hover, Matterport, OpenSpace.Ai), and energy management (e.g. Readaptive, BlocPower, Span.io). These companies all got over $100 million each in funding.

The chapter next outlines the relationship between Industry 4.0 and Construction 4.0, with examples of construction systems, machinery and equipment available in 2022. Those examples show the fourth industrial revolution spilling over into many aspects of construction products and processes, as technologies get commercialised. However, rather than a general discussion of what Construction 4.0, smart sites or cyber-physical systems might look like (all of which are becoming widely available) the chapter discusses the development of a digital production system where the three dimensions of industry development are revisited. A discussion of construction of the built environment and AI follows, then the role and potential importance of digital fabrication and 3D concrete printing is considered.

Construction 4.0 Technologies

With the idea of Industry 4.0 came Construction 4.0, with the same problem of mixing enabling technologies like the IoT and AI that are widely applicable, with task specific industry requirements. A 2020 book called *Construction 4.0* defined it as 'a framework that is a confluence and convergence' of three types of technologies: industrial production with prefabrication, 3D printing and OSM; cyber-physical systems are actuators, sensors, IoT, robots, and drones; and digital and computing technologies include BIM, video and laser scanning, AI and cloud computing, big data and data analytics, reality capture, Blockchain, simulation, augmented reality, data standards and interoperability, and vertical and horizontal integration.[190] That is a comprehensive list of twenty-first century technologies ranging from already mature to early stage development, although it is incomplete in missing advances in materials like engineered wood factories, higher performing concrete additives and sealants, 4D printing of reactive and shape memory materials, and roller press printing of smart fabrics (for both facades and interiors).

Rearranging the order of this jumbled list of technologies from their types to their level of development shows how close to wider use many of them are. Technologies that are currently used are prefabrication, OSM, building information modelling (BIM), data standards and interoperability, and other technologies that are commercially available are 3D printing, drones, video and laser scanning, cloud computing, simulation and augmented reality. Technologies that are under development are actuators, sensors, IoT, robots, AI, big data and data analytics, reality capture, and Blockchain.

These technologies are no longer in the future, they are being rapidly commercialised by both new entrants and incumbents. Some like exoskeletons are still in early-stage development, but others like unmanned aerial vehicles (UAVs, or drones) are already spreading throughout industry. Drone monitoring of onsite progress, with real time matching of scans and pictures with the project BIM model or digital twin, is available with systems like those from Skydio and Skycatch with UAVs, or Icon's tracked lidar scanner. Trimble mounted a laser scanner on a Bostin Dynamics mobile Spot robot for automated site analysis, in a system that became available in late 2021. In fact, by then, momentum had built behind a wave of product launches for the new machinery, equipment and systems needed for Construction 4.0. The table below has industry 4.0 technologies for construction products and processes, it shows the range of technologies involved and gives examples of companies that are providing them and is an outline of the boundaries of the construction technological frontier in 2022.

Table 7. Examples of companies with construction 4.0 technologies

Product or service	Company
3DCP with boom system	Mighty Buildings, ICON, Aris, 3D Constructor
3DCP with gantry system	COBOD, BIG, Black Buffalo, Contour Crafting
3DCP with mobile robot	Imprimere, Printstones
3D metal printing	HP, GE, MX3D, Aurora Labs, Arup, 3D Metalforge

3D fibre reinforced polymer	Branch Technology (facades)
Site layout printers	Rugged Robotics, Dusty Robotics
Autonomous equipment	Build skidsteer, Hilti Jaibot (for anchor points)
Bricklaying robot	FBR Robotics 'Wall as a service' (truck mounted system) SAM (tracked system)
Cloud based digital twin	VectorWorks, Dronebase, Hover, Airware, Versatile, Digital Construction Works
Design and fabrication platforms	Juno, Modulous, PT Blink, Project Frog, Hypar, Plant Prefab
Construction site IoT	Trimble, Pillar Technologies, T3,
Drone monitoring tracked	Icon, Trimble with BD Spot
Drone monitoring UAV	Skycatch, Vinci, Skydio
Exoskeletons	Esko, HULK, Japet, Sarcos
IoT linked building sensors	AmbiMate, IBM, Legrand, Honeywell, Panasonic, Siemens
Remote controlled equipment	CAT, Komatsu, Brokk demolition robot

Smart helmets	Trimble Hololens, Daqri
Plasterboard robots	Canvas, OKIBO

The problem with the concept of Industry 4.0 in general, and Construction 4.0 in particular, is the large number of technologies included, all of which have their own development paths and many of which are enabling technologies, like IoT and VR. In the same way as the concept of an industrial revolution by itself is not an insight into the inventions and innovations involved, Industry 4.0 is not an explanation of what is happening to industries and firms. Some Industry 4.0 technologies like BIM and platforms are in already use, some like 3D printed concrete are moving out of the experiment stage to commercialisation, others like smart contracts are still conceptual. For OSM there are automated engineered wood factories, the UK government is funding a construction manufacturing hub, and the Autodesk BUILD Space in Boston is an innovation centre.

It was at the first public demonstration of virtual reality (VR) headsets in 1990 that William Gibson made his now famous observation 'The future is already here, it's just not very evenly distributed.'[191] Those early, primitive, nausea-inducing systems were clunky and expensive, but after a couple of decades of development the costs of the key components, particularly small high-resolution screens and sensors, had fallen to the point where consumer products were possible. Released in 2016 were VR headsets from Oculus, Microsoft and Samsung, and in 2017 everyone from architects to zookeepers started thinking and talking about how this 'new' technology could be used. Smart helmets arrived, and the Microsoft HoloLens 2 was released in February 2019 connected to a BIM model running on Trimble software to provide an AR view of the project. By 2020 drones were coming into widespread use. Komatsu and Vinci had automated earth moving equipment using sensors and drones linking real-time data about site works to remote operators, using BIM to monitor progress. Volvo and Otto had automated trucks and Rio Tinto's mines had driverless trucks and trains linked to a control centre. The SAP Connected Construction system uses the IoT to link equipment to a remote controller.

Because the concept of Construction 4.0 includes so many different technologies it lacks focus. Components such as sensors, scanners and actuators in the future might be combined into one system, others like the IoT are generic and construction will adopt them as applications develop. Technologies like drone monitoring will clearly be important. However, only a few technologies have the potential to affect construction of the built environment in a fundamental way.[192] Blockchains and smart contracts might in the future have such an effect. At the present moment two important new technologies are AI and digital fabrication, because both are entering the stage of their life cycle when use can spread widely and quickly, after many years of development.

AI in Construction

BIM can run on platforms, it allows access to cloud manufacturing, it is being combined with virtual reality (VR) and augmented reality (AR) systems for a holographic 3D virtual project that contains the details of a building, and that information can be shared through a project management platform with all project participants. At this point the expectation is that VR will be used more on the design side by architects, planners and engineers, while AR will have a larger footprint on construction sites, although some construction firms have started looking at using VR in areas like safety and training. BIM is obviously central to these technologies. Other uses include drones measuring earthworks and monitoring projects by matching site work to BIM plans for buildings and excavators.

If there is a technology that ties this disparate collection of Industry 4.0 digital applications together it is AI, and all of them will generate data. Cloud based platforms providing data for analytics and machine learning will underpin the machine learning needed for construction AI. If software platforms that provide cheap, standardised business processes become widely used they will allow training of machine intelligence on ever larger data sets. Even without major breakthroughs, this means the role of AI can develop along with the applications for its use over the next decade or so. Therefore, AI use can steadily, if perhaps slowly, spread through a diverse system of production like construction.[193]

In 2016 the AlphaGo system developed by DeepMind (now Google) defeated the Go world champion, by using multiple machine learning algorithms for training and a sophisticated search procedure while playing. This sparked a surge in both interest and investment in big data, pattern matching and machine learning across the five general categories associated with intelligence: logical reasoning, knowledge representation, planning and navigation, natural language processing, and perception.[194]

Current AI technology already provides services such as GPS navigation and trip planning, spam filters, language recognition and translation, credit checks and fraud alerts, book and music recommendations, and energy management systems. It is being used in law, transport, education, healthcare and security, and for engineering, economic and scientific modelling. Advanced manufacturing is highly automated and, as expected with such new technologies, there are many new applications under development.[195]

The linking of computers to a widening and increasing array of sensors and input devices is generating what appears to be intelligent behaviour by these machines, as they learn to process big data sets and react to changes in the state of the world. Examples are predictive maintenance of equipment, managing HVAC systems, translating languages, diagnosing x-rays and driving cars. This form of machine intelligence is known as 'narrow' AI. To accomplish a task machines are 'trained' on data sets to improve their performance, but they cannot transfer that training and learning from one task to another, nor explain or understand what has been learnt.

The application of narrow AI and machine learning to data associated with construction can be done for the operation of plant and equipment, the progress of a construction project or the condition of a building.[196] The increasing capabilities of data processing and storage will make machines more intelligent, in that they can manage such limited, defined tasks well. Over time these intelligent machines will deliver information that can help people manage large and complex data sets with comprehensive data visualization and application tools. The machines become 'smarter' as they acquire more data, capabilities and sensors, and help people navigate complicated processes and assist them in making decisions. Because the data is available, predictive maintenance of plant and equipment using AI is an early application. In the various forms that AI takes on its way to the construction site, in one way or another it will become central to many of the tasks and activities involved.

Another example is the use of AI for generative design. These systems use data on planning and building codes to provide numerous design solution and generate alternatives. The system coordinates design objectives with project materials, construction and manufacturing costs and methods. Based on the objectives and parameters, a generative design AI quickly produces design options, each of which could take an architect days to do. The system also prevents plans for the mechanical, electrical and plumbing systems from clashing, learning from each iteration until it comes up with the ideal model. Another AI could then coordinate offsite construction with product specification, fabrication and delivery. With 3D printing technology, complex design options can be produced.

Once construction has started, the integration of real-time data with the model in the digital twin could be the core role of AI in onsite construction, optimizing outcomes by predicting potential points of failure in a plan.[197] The level of AI in Table 3 envisages such capabilities for a group of intelligent machines that have been individually trained to collect and manage data from the stages of a construction project. Outsourced business processes can provide the data needed for training intelligent machines that are supervised by users and help them manage complicated processes.

An AI acting as a project data manager could integrate the data from many sources to continually update a project's schedule, work plan and cost estimates, matching progress and performance to iterate those plans for the project's managers. This AI assists users' decision-making by generating and evaluating options. Such a system would be operated by a voice activated interface, with the progress updates included and access to expert systems for specialist areas provided. It would generate design options and provide full visualisation of a shared building model linked to the schedule and site work plan. There would be real-time supply chain data on fabrication and logistics through cloud-based platforms. An AI can iterate and optimise the schedule and cost plans for a project, based on that data, allowing the project management team to match performance with plans, in real-time, for every aspect of a project.

Optimizing with AI is done through prediction of the effects of current circumstances on a plan's time and cost forecasts. It does this by combining, managing and integrating data from many sources with analytical and machine learning capabilities to allow reliable predictions about the state of the world, such as the operating status of plant and equipment, the daily progress of a construction project or the energy consumption of a building. Although there are many difficulties and technical challenges involved in reaching that level of machine intelligence, it is not unrealistic to think many of them will be solved in the next decade as the capabilities of deep learning and cloud-computing increase. However, very large data sets to cover the range and diversity of construction projects will be required for machine learning.

Table 8. Applications of machine intelligence in building and construction

Stage	Intelligent machines Users have supervisory role	Artificial Intelligence Users make final decisions
Design	Coordinate design and consultants Manage design and component libraries, building standards and planning codes	Generate design alternatives Maintain BIM model Iterate design schedule
Fabricate	Coordinate design and production Manage production data from suppliers and fabricators	Component data analytics Iterate production schedule Manage storage and delivery

Project Management	Coordinate contractors, fabricators and suppliers Manage logistics and site	Data integration and visualisation Iterate cost and work plans
Construct	Link model and site data Coordinate and monitor site machinery and equipment Scan and survey work done	Data visualisation and management Iterate site workplan
Operate	Manage environmental conditions Monitor structure and components	Maintain digital twin Iterate maintenance plans

Note: The data required for the coordination and management role of intelligent machines can come from widespread use of standardised, outsourced cloud-based business processes. That data then becomes a series of training sets needed for deep learning, the current level of AI technology.

Platforms use forms of AI to monitor and manage the data they produce, the function of intelligent machines. Examples are LinkedIn (matching jobs and people), Skype (simultaneous translation of video calls), AWS and other cloud-computing providers, and marketing, legal and accounting software systems. As digital platforms providing building design, component and module specification, fabrication, logistics and delivery become widely used, they could also provide outsourced business processes. These processes are usually cheap, because they are standardised, and are available to large and small firms. Cheap, outsourced, cloud-based business processes can lower fixed costs and thus firm size, because firms can focus on their core competency and purchases services as necessary as they scale, with lower costs leading to more entry and more innovation. If these digitised business processes are cost-effective and become widely used, they can provide much of the data needed to train machines, so as more projects are completed the data from their digital twins can add to the data set.

One problem with AI is the 'black box' effect, explaining predictions can be difficult, particularly when they would not normally be considered by users. However, if interactions between artificial and human intelligence are situations where there are many possible solutions and no one knows the best one, AI can help with problem solving and optimising production models by identifying optimal solutions, while humans help AI by providing knowledge these systems lack. Using the trial-and-error process traditionally followed in construction, integration of AI in the supply chain for design and fabrication can converge with other Industry 4.0 technologies for project delivery.

3D Concrete Printing, Digital Fabrication and Onsite Production

There are three methods for 3D printing: stereolithography, patented in 1986: fused deposition modelling, patented in 1989: and selective laser sintering, patented in 1992. It didn't take long before research into 3D concrete printing (3DCP) began, focused on developing the equipment needed and the performance of the materials used. Twenty years later there were over a dozen experimental prototypes built, extensively documented in the 2019 book *3D Concrete Printing Technology: Construction and building applications*,[198] which also has details on the materials science required to identify successful mixtures and admixtures. The information needed to create a 3D blueprint is generated during design, and it is a relatively small step to move from a BIM model to instructions for a 3D printer.

By 2022 the commercialisation of 3DCP was underway, with two types of systems available. One using a robotic arm to move the print head over a small area, intended to produce structural elements and precast components, the other a gantry system for printing large components, walls and structures. The Additive Manufacturing Marketplace had 34 concrete printing machines listed, ranging from desktop printers to large track mounted gantry systems that can print three or four story buildings. Companies making these machines are mainly from the US and Europe, and Table 9 also has details on the type and size of a selection of machines. There are also several companies offering 3DCP as a service at an hourly or daily rate.[199]

One 3DCP company is Black Buffalo, a subsidiary of South Korea's Hyundai group based in New York. Their NexCon gantry system takes around 11 hours to build and eight hours to take down. Using a proprietary ink developed over a few years of research involving a lot of trial and error (and getting approval for building codes), the machine can print up to four stories with a crew of five people. One person is required to monitor the nozzle and insert stiffening frames every few layers to provide structural strength, the pump needs two people and a helper, plus a site foreman or engineer. As well as walls it will print floors and precast elements. Black Buffalo expects to sell over 100 of these printers in 2022-23, and they are available for rent at $1,000 a day.

Table 9. Some companies making 3D concrete printers

Company **Country**	**Machine and** **Size in meters**	**Type**	**Cost in US Dollars**
Black Buffalo *United States*	NexCon 1-1 3 story 4 story	Gantry	$400,000 $750,000
COBOD *Denmark*	Bod 2 14.6 x 50.5 x 8.1	Gantry	$200,000 to $1m
Imprimere *Switzerland*	BIG 3D 5.7 x 6.0 x 6.2	Gantry. Prints large components	$1,757,000

ICON *United States*	Olympus 2.6 x 8.5 x 2.6	Gantry	$150,000
Constructions 3D *France*	Maxi Printer 12.2 x 12.2 x 7	Mobile robotic arm on 4 legs	$495,000
Luyten *Australia*	Platypus 6 x 12 x ...	Gantry	$36,000
PrintStones *Austria*	Baubot	Mobile robotic arm on tracks	$150,000
Massive Dimension *United States*	MDPC 2 x 2 x 2	Fixed robotic arm	$80,000
CyBe Construction *Netherlands*	CyBe RC 3Dp 2.75 x 2.75 x 2.75	Fixed robotic arm	$205,000
MudBots *United States*	MudBot 3D Up to 22 x 22 x 15	Gantry	$35,000+
WASP *Italy*	DeltaWASP 1 x 1 x ...	Robotic arm	$100,000+

Source: Additive Manufacturing Marketplace, April 2022.

Concrete printing is only one part of the development of additive manufacturing. In mid-2022 the Additive Manufacturing Marketplace listed 2,372 different 3D printing machines from 1,254 brands. The number of printers and materials used were: 364 metal; 355 photopolymers; 74 ceramic; 61 organic; 34 concrete; 24 clay; 20 silicone; 19 wax; and 19 continuous fibres. Many of these printers could be used to produce fixtures and fittings for buildings. Producing components onsite from bags of mixture avoids the cost of handling and transport, and for large items avoids the load limits on roads and trucks. There are also printing services and additive manufacturing marketplaces being set up. These link designers to producers with the materials science, specialised equipment and print farms capable of large production runs and manufacture on demand. Examples are Dassault Systems 3DExperience, Craft Cloud, Xometry, Shapeways, 3D Metalforge, Stratasys and Materialise.

Additive manufacturing is a major part of a broader system of production known as digital fabrication. Neil Gershenfeld described digital fabrication as turning 'bits into atoms and atoms into bits' using fabrication laboratories (fabs) producing 'assemblers' that provide the cutters, printers, millers, moulders, scanners and computers needed for designing, producing and reproducing objects.[200] These tools include traditional subtractive ones for cutting, grinding or milling, but the focus has been on research into new methods of additive manufacturing using different methods of layering materials using 3D printers. Printing of metal, ceramic and plastic objects from online design databases is spreading from hobbyists and initial users to industry applications and wider acceptance.

Gershenfeld, who founded the first fab in 2003, defined digital fabrication, as 'the seamless conversion of design and engineering data into fabrication code for digitally controlled tools.' The definition in the *Construction 4.0* book is less focused, a 'method or system which relies on digital fabrication entirely or to a significant degree, either in prefabrication or on-site construction. Examples of digital fabrication processes in construction include robotic fabrication and assembly, large scale additive manufacturing, and the use of specialized automation systems for material processing in areas ranging from advanced fabrication of metal or timber assemblies to various forms of concrete processing to the fabrication of multi-material composites'.[201] The chapter on digital fabrication has over a dozen examples, with many of the projects shared with other chapters in the book on 3D concrete printing, robotics and automation.

At is broadest, digital fabrication is any means of turning design information into physical products using automated processes. There is a well-established global maker movement behind the growth of digital fabrication. In 2009 the Fab Foundation was established at MIT as a non-profit with annual conferences and providing educations and training.[202].

The Foundation coordinates an international network of 1,500 fabrication laboratories (fabs) in 90 countries, many in university design and architecture schools, and the 'FABLAB Movement' is an even broader collaborative effort that includes hobbyists and tinkerers working on digital design and fabrication code.[203] This network takes existing technologies used in fabrication like cutting, milling and rolling done by numerically controlled machines, which have been around for decades, but uses them for design, which is new. The digital linking of design to fabrication is the beginning of another stage of development. The World Economic Forum also has a network of 14 Centres for the Fourth Industrial Revolution, and Autodesk has three BUILDSpace laboratories. The 2017 UK industrial strategy included funding for a manufacturing hub and along with aerospace and automobiles targeted construction, 3D concrete printing and OSM.

Gershenfeld argues digital fabrication will follow a similar exponential development path as Moore's Law for transistors, with the number of fabs doubling every two years and their cost halving. This would make local production of many objects and items possible.[204] Gershenfeld suggests the technology is now ready to become widespread, and is at a similar stage to PCs in the early 1990s:

> Digital fabrication shares some, but not all, of the attributes of communication and computation. In the first two digital revolutions bits changed atoms indirectly (by creating new capabilities and behaviours); in the third digital revolution, the bits will enable people to directly change the atoms .... Across the global network of fab labs, we can already see a steady stream of innovations around cost-effective models for individuals and communities to make clothing, toys, computers and even houses through designs sourced globally but fabricated locally.[205]

Digital fabrication is at or close to the tipping point, as its use extends from hobby to experimentation and industry adoption. Just as concrete in the early 1800s moved from the fringe toward the centre of construction as the underlying technology and associated equipment improved, fabs can follow a similar path over the next decades of the twenty-first century. Although *Construction 4.0* concluded 'Today, industry adoption of digital fabrication is still very limited and is not deployed at scale in the industry',[206] the technology is advancing rapidly and many demonstration 3D concrete printing projects have been completed successfully. In a 2020 report that has many examples of current use, ARUP argues: 'The opportunities unfolding with digital fabrication not only demonstrate new techniques in full-scale pavilion fabrication, but also provide new methods to solve design, business and societal challenges.'[207] Arup is one of a number of specialized consultants providing digital twins and design to fabrication capability on projects.

ONSITE AND NEARSITE PRODUCTION WITH DIGITAL FABRICATION

Digital fabrication is a technology whose use has a high probability of becoming ubiquitous. In construction, the focus so far has been on 3D printing of concrete, with experimental systems by the early 2000s,[208] and by 2019 there were over a dozen examples of buildings completed using the technology.[209] However, the potential of 3D printing in construction is not limited to concrete. Once a BIM model of a project has been created it can be used to provide instructions for production of both structural and decorative components of a building. Mobile digital fabs in shipping containers can produce some of those components onsite. Local firms offering manufacturing on demand from print farms can produce large runs or specialised components, a nearsite form of production rather than OSM. The combination of digital twins and digital fabrication would be transformational if it allows onsite and nearsite production of some or many building components, by fundamentally altering existing economies of scale in the industry. As well as 3D concrete printing, other materials like steel and plastic can be used to make components and fittings on or near the building site. A modular fab in a container customised for construction, or even a specific construction project, can be set up onsite to produce components as the schedule requires. Large sites might need a fleet of fabs. Restorations and repairs can be done with replacement parts made onsite from scans of the original.

Mass production will always have a role, but market niches currently occupied by some or many manufacturing firms may be replaced by new production technologies based on BIM, linking localised digital fabrication facilities with online design databases. Combined with robotic and automated machinery and assemblers, digital fabrication and standardised parts opens up many possibilities. Adding new materials to the 3D palette through molecular design and engineering, or upgraded versions of existing materials, may unlock other unforeseen design and performance options.

If this eventuates some, possibly a great deal, of the current construction supply chain based on mass-production of standardised components will become redundant. For example, an onsite or nearby fab with printers and moulders might produce many of the metal, plastic and ceramic fittings and fixtures for a building during its construction. The digital twin of the project, which might be outsourced, can link the design and fabrication stages to the site and the project. Digital fabrication produces components and modules designed to be integrated with onsite preparatory work and assembled to meet strict tolerances. Project management would become more focused on information management, and the primary role of a construction contractor might evolve into managing this new combination of site preparation work and integration of the building or structure with components and modules, some of which may be produced onsite in a fab if economies of scale permit. The strength of this effect will be determined by those economies of scale. Beyond site preparation other site processes may be restructured around components and modules that are designed to be assembled in a particular way, and machines to assemble those components and modules can be fabricated for that purpose.

If onsite and nearsite production becomes steadily cheaper the industry would, perhaps slowly, reorganise around firms that best manage onsite and offsite production and integration of digitally fabricated parts. Contracting firms would become more vertically integrated if they are fabricators as well, reinventing a business model from the past when large general contractors often had their own carpentry workshops, brick pits, glass works and so on. With outsourced business processes and standardized site and structural work, that fabrication and integration capability would be a key competitive advantage of a construction firm. Firms will be integrating automated production of components with design and construction using offsite manufacturing and onsite fabrication, using platforms that coordinate building design and specification with manufacturing, delivery and onsite assembly. Open platforms will be like new digital ecosystems. Closed, internal platforms will be developed by larger, vertically integrated firms with the resources to manage the system.

Industrialized materials like concrete, steel and glass affected the organization of onsite processes as they were improved with incremental innovations. The development of digital fabrication should follow a similar path to concrete, where over decades the machinery (mixers, pumps), processes (formwork systems) and materials (reinforcement, concrete strength, setting agents) were developed. Growth in digital technologies is faster than analogue, so instead of the many decades of innovation taken for concrete technology to develop, it might take a decade for digital fabrication to become cost effective if the cost of fabs falls and the supply chain of raw materials develops as it did for personal computers in the 1990s. Contractors would become more vertically integrated as they also become fabricators, managing a combination of onsite and nearsite production to deliver projects.[210]

Dimensions of Digital Construction

In the nineteenth century technological changes in construction operated over the three dimensions of industrialization of production, mechanization of work, and organization of projects. The fourth industrial revolution can be expected to also work along these dimensions as it reconfigures them by linking data and participants through the life of a project. As cloud-based digital twins and software-as-a-service become more widely available, the digitization of construction will acquire further momentum.

Table 10. Dimensions of development for Construction 4.0

Dimensions of development	Construction and the fourth industrial revolution: Possible developments
Production of components and materials	Platforms integrate design and production with voice-controlled 3D models of buildings and components. Generative design used for selection of components and modules from online design libraries, both open-source and private.

	AI monitors compliance with building standards and codes. Developments in digital fabrication, design software and molecular engineering allow a range of new production technologies to spread through the industry. Economies of scale for onsite versus offsite production determine what, where and when components are produced.
Mechanization and automation of tasks	Site workers have exoskeletons and smart helmets available. Many onsite tasks can be done by teams of robots. Heavy machinery and equipment is operated remotely, often with some autonomy for repetitive tasks. Assemblers can be designed and fabricated to install components and modules, which can be designed to be handled by those assemblers.
Organization of projects	Cloud based platforms integrate delivery of the physical project with its digital twin, with real-time data and monitoring of activities and tasks. Standardised, outsourced cloud-based business processes used. Contractors become more vertically integrated and focus on integration of site work, managing OSM and onsite production. As built digital twins required by clients.

The construction technological system is extraordinarily wide and diverse, and the various parts of the digital production system are in various stages of development. There are very many possible futures that could unfold over the next few decades. However, machines that can use data and information to both interact with each other and work with humans are moving from operating in controlled environments like car manufacturing to unpredictable environments, like driving a tractor or truck. In many cases, like remote trucks and trains on mining sites, the operations are run as a partnership between humans and machines, as the saying has it 'running with the machines not against them'.[211]

A 2020 US survey asked firms about their use of advanced business technologies. Cognitive technologies that help machines to 'perceive, analyse, determine response and act appropriately in [their] environment' (a standard definition of AI), included machine learning, machine vision, natural language processing, and voice recognition software. Robotics (i.e., automatically controlled, reprogrammable, and multipurpose machines), radio frequency identification (RFID), touchscreens/kiosks for customer interface, automated storage and retrieval systems, and automated guided vehicles were also included. Large firms invested in and used more of these technologies than small firms. In both Construction and Professional services the most widely used technologies were touchscreens, voice recognition and machine learning.[212]

BIM as Industry Policy

The introduction of international standards for BIM and digital twins in 2019 may be the milestone that marks the transition to a digitised system of production.[213] The ISO 19650 standards provide a framework for creating, managing and sharing data on built assets, an essential step because, in a technocratic system like construction, standards play a key role in establishing consensus on what is to be done and how.

For built environment industries, BIM has been promoted as the solution to the problems of poor documentation, fragmentation and lack of collaboration in building and construction for many years. BIM had its origins in 1960s 2D drawing programs that developed into architectural drawing software. Two companies dominate the market, Autodesk was founded in 1982 and Bentley Systems in 1984. The first version of ArchiCAD's file exchange solution was released in 1997, which allowed multiple designers to work on a collaborative platform. At this point enthusiasts began believing in BIM as a universal panacea for the problems and issues endemic to construction. Twenty-five years later they are still waiting, despite the fact that BIM is no longer a new technology but an application widely used in construction, one that is now offered as a cloud-based software-as-a-service to manage and maintain project digital twins.

BIM has not, however, been disruptive as we understand the idea, at least not so far. This is not a new technology. Therefore, BIM does not qualify as transformative, rather it is the required enabler of further developments, a necessary foundation for the transition to the construction technological system in the digital age. BIM is more like digital plumbing underpinning digital construction than an elevator to higher performance.

BIM is plumbing because the digitized data it generates gets shared across the network of different built environment industries. At a basic level this is just sharing files and managing documentation. However, BIM can run on platforms, it allows access to cloud manufacturing, it is being combined with virtual reality (VR) and augmented reality (AR) systems for a holographic 3D virtual project that contains every detail of a building, and that information can be shared through a project management platform with all project participants. At this point the expectation is that VR will be used more on the design side by architects, planners and engineers, while AR will have a larger footprint on construction sites, although some construction firms have started looking at using VR in areas like safety and training. BIM is obviously central to these technologies. Other uses include drones matching site work to BIM plans for buildings and excavators measuring earthworks. Some clients are demanding as-built digital twins to manage their buildings with.

These is little practical difference between a country's industry policy and national industrial strategy. They are both typically framed around competitiveness and productivity, focus on innovation and R&D, and follow pathways and roadmaps through scenarios and scoping studies. Some industries like agriculture, steel and automobiles are regarded as strategic and have always been surrounded by rules and regulations and subject to government intervention. Governments' have science and technology policies that influence industrial structure and macroeconomic policies that affect economic development. For many countries the emphasis in industry policy has shifted to industry 4.0 technologies and AI, as governments and industry respond to these technologies.

Countries have taken different approaches to promoting BIM. Broadly, Scandinavian and western European countries, Singapore and the UK followed a government-driven approach, but Australia and the United States (US) a more industry-driven approach. However, the US General Services Administration (GSA) established the first public sector program in 2003, the National 3D-4D-BIM Program, on best practices for design and construction teams. The GSA was also the first client to require mandatory use of BIM in 2007, for program verification. The first government BIM roadmap was from Singapore, for 2010-2015, by the Building and Construction Authority, with a second in 2016 that included BIM for facility and asset management and the *BIM for DfMA Essential Guide* for integrating BIM and DfMA[214]. BIM use in Singaporean construction is among the highest worldwide.

The UK Government Construction Strategy 2011–2015 mandated fully collaborative 3D BIM for all public projects by 2016. Importantly, the UK also began publishing BIM standards to provide guidance for industry on how to produce, exchange and use information in BIM. France initiated its 'plan de transition numérique du bâtiment' in 2014 with significant investment in digital technologies, followed by a roadmap in 2015, a BIM guide for building owners in 2016, and a standardisation strategy in 2017. By 2020 most western and northern European countries had plans to mandate BIM in some way, although the level of use varied greatly between countries, with BIM adoption in the UK, Denmark, Germany and France similar to the US, Canada and Singapore, but Southern European use much lower.

In the US many land use and building codes are local, and a range of different approaches have been followed. The US also has widely used standards and guides from both government and industry bodies. The GSA 2009 Guides were on 3D imaging and 4D schedule management, extended to life-cycle management in 2011. The American Institute of Architects published six series of guidelines after 2007 for the use of BIM in the design and operations of projects for architects. The National BIM Standard was published in 2009, updated in 2012, and is in its third version. The US followed an industry-driven approach and, compared to Singapore and the UK with their BIM mandates, the government was less involved.

In Australia, the Commonwealth Government released a national BIM initiative in 2012 and recommended requiring full 3D collaborative BIM for all Australian government projects by 2016. However, with no mandates or targets for use nothing actually happened. As in the US, policies and uptake varies across the states. In 2018 the Queensland government started mandating BIM, to be expanded to all built assets by 2023.[215] Other states are following.

Industry has a collective action problem because the cost of adopting a new technology is significant and skills are typically in short supply. Firms will invest in BIM if they believe that they will profit by it, but legitimately fear future technical progress could make today's investments unprofitable as change makes today's technologies obsolete. Paradoxically, when innovation and technological progress is rapid uncertainty may hold back investment by firms because there may be a better, cheaper technology available tomorrow. Why invest today if there will be a competing technology that is half the price in a few years' time?

ROADMAPS AND BIM MANDATES

Two common but erroneous explanations for why BIM is not widely used are the inertia of a conservative industry culture and the incremental process followed by clients in requiring BIM. In fact, BIM use across professional services, contractors, subcontractors and clients is very unevenly distributed and varies greatly between large and small firms. Governments around the world have been promoting BIM, and these individual government policies also vary greatly. These policies broadly have roadmaps with stages for BIM adoption, using both level of use and size of project as targets, and are intended to allow time for industry to adjust.[216] The UK is a good example of the policy approach being followed by governments to increase use of BIM. With a BIM mandate introduced in 2011, a decade later the UK had a high level of use of BIM. A 2021 survey by the UK BIM Alliance[217] with over 1,100 respondents found 65% were implementing BIM and used it on around half their projects and 30% were using ISO 19650.

With the mandate the government provided clauses covering contentious issues in construction contracts (such as intellectual property and data ownership) that worked with rather than against industry practice and culture. In 2015 standards BS 8541-5 and 6 on offsite construction and modular buildings were released. The Construction Strategy was extended to 2016–2020, with a single shared building model to be held in a centralized repository for operation of assets over their life cycle.[218] The 2018 BIM Framework[219] based on ISO 19650 provided a roadmap for the firms and clients and the development of standards provided a toolkit.

A small number of countries have implemented national BIM mandates:[220]

2004 Singapore for public construction projects
2007 Finland for all public projects over 1 million euros
2007 US General Service Administration and the Army Corps of Engineers required use
2010 South Korea public construction over KRW 500 million from 2016
2011 UK for public building
2018 Spain for public construction
2019 Abu Dhabi for all major projects
2020 Germany for Federal infrastructure projects

Many other countries have published roadmaps, standards and guidelines since 2015 without so far following up with a mandate, for example Austria, Australia, France, Switzerland and Japan are at this stage. In every case the underlying assumption is that BIM will become business as usual over the decade of the 2020s, but at the beginning of the decade countries that were early movers like Singapore, Finland and the UK have the highest use of BIM. There are also state mandates in the US and Australia: Wisconsin required BIM for projects over $5 million in 2010 and Queensland for public projects in 2018. In the early 2020s major projects for both public and private clients worldwide are done with BIM.

The UK construction strategy applied to all firms involved in projects, and thus included designers, consultants, suppliers and subcontractors as well as the contractors who had been the focus of earlier construction industry policies. Also, the new strategy targeted technology adoption not the separate industries of residential building, non-residential building and engineering construction. The differences in the subcultures of these separate industries accounts for the differing rates of uptake of BIM found across firms in the UK since the launch of the strategy. Also, national and local governments, universities, regulators and industry bodies were all given significant but loosely specified roles in these policies to support industry engagement. Some clients are demanding as-built digital twins to manage their buildings with after completion.

Achieving industry policy goals requires a great deal of coordination, determination and long-term commitment, qualities not always associated with government policy. Over the decade after the UK government launched a new Industry Strategy in 2011 and the Construction Industry Strategy in 2015 there was investment in capability, new standards were developed, and BIM requirements increased usage. This new form of industry policy was more about collaboration between the public and private sectors,[221] rather than imposing unrealistic outcomes on industry. Industry policies do not have to be original or innovative to be useful and effective, as the success of the UK and other countries with BIM mandates in promoting use of BIM shows.

6 The Built Environment and Industry Policies

Responsibility for public sponsorship generally was somewhat elusive, lying along attenuated administrative chains
Christopher Powell[222]

In a time of rapid urbanisation and great social and environmental challenges, the built environment and associated housing, infrastructure and urban policies have become central issues in public policy. The quality of the built environment is a major determinant of the quality of life. Further, cities are at the centre of the modern economy and, in a fundamental sense, how well cities function depends on how well the many and diverse industries, firms and organizations across the built environment sector can design, deliver and operate the projects required. The resilience of cities to climate change is being tested as temperatures increase and fires and floods become more intense. However, because of the range and complexity of these issues it is difficult for governments to develop and implement coordinated built environment industry policies that address these issues satisfactorily.

Industry policy was out of favour for a couple of decades before the financial crisis in 2007-08 in the US, UK and Australia, although the European Union (EU) and many Asian countries followed well developed national strategic plans. This was partly ideological, a view that policy is another government economic intervention that requires picking winners, and partly because some issues traditionally addressed by industry policy like tariffs and market access moved into negotiations around trade policy, at both the global level and in the increasing number of regional and bilateral trade agreements.

Following the financial crisis governments looking for sources of economic growth and employment creation began focusing on specific sectors in manufacturing and services where they saw opportunity in global value chains.[223] Environmental standards and policies supporting renewable energy were developed.[224] Industries like pharmaceuticals and biotechnology, semiconductors, aerospace, IT, AI, cars and steel have featured in industry policies in many countries. Any policy intervention intended to strengthen the economy is an industry policy, and governments regularly establish priorities and target industries.[225] Countries protect or favour industries with legislation for many reasons but some of them are strategic and long term, like innovation programs with their associated challenges, roadmaps and milestones, and many of these programs currently involve digitization and automation in some form.[226]

Government policies targeting supply side issues are not as high profile as others, they don't get regular updates like monthly unemployment or quarterly GDP statistics and capture attention like announcements of interest rate changes. Because productivity has become the measure used for industry performance, despite the statistical questions that raises, it has often been the target for government policy. However, many of these policy measures will only affect productivity in the long run, examples are education, training, infrastructure, innovation, R&D, capital expenditure subsidies, and pilot or demonstration projects. Therefore, results take time and thus take longer than the electoral cycle to develop, so there is often little benefit to the government of the day even if a policy is working well.

When the intention of such policies is to influence a country's economic structure and industry development they can be described as industrial strategy or industry policy.[227] What history generally does show is that it is hard to get an industry strategy right and implementation is difficult. Traditionally manufacturing was the focus for industry policy, but after 2007 the approach became more about coordinating a wide range of policies to achieve both economic and social objectives.[228] Climate change and environmental issues have become a focus for a range of industry policies aimed at reducing emissions.[229] The rollout of protective equipment and vaccines during the Covid pandemic in 2020-21 both tested and accelerated this new approach.

As well as common industry policies targeting innovation, training or business investment, construction of the built environment is also subject to many other government regulations, legislation and policies. On the demand side interest rates, taxes, public infrastructure spending, urban development and housing policies are all important, but are also external to the built environment sector itself and are determined by a wide range of factors beyond the sector. Then there are the effects of planning and environmental regulations and restrictions limiting the supply of new housing or infrastructure, an issue that has featured in recent debates in many countries and spills over into other issues around the affordability of housing and the location and cost of major projects. The number of different government departments and agencies involved in regulating the built environment is often a major barrier to innovation because coordination is difficult and there are many opportunities for incumbents to delay or derail progress when reforms are proposed.

The public sector in many countries is collectively the largest client for construction, but the expenditure is spread over departments like health, education, transport and defence, and there is unrelenting pressure from public sector clients for the lowest possible cost of work. In practice, there are significant institutional constraints on government buying power. Although reports in many countries have recommended leveraging public procurement of buildings and structures to push industry reform this is not widely used, despite being common practice in Asian countries like Singapore and Japan.[230]

While it is a fact that governments can have major impacts through regulation, tax, innovation and R&D policies, their effect is uneven and hard to discern. For example, large firms in capital intensive industries like cement respond to industry policies differently to large contractors, as do professional service SMEs compared to construction trade SMEs. There are, however, two areas where governments have had some success in promoting industry development: BIM mandates and building standards and codes.

From Reforming Construction to Mandating BIM

Contractual relationships were the focus of much of the reform agenda to improve industry performance. In the UK the Simon Committee report in 1944 on building contracts had called for cultural change, as did Latham 50 years later. Egan introduced benchmarking against best practice to improve productivity and Constructing Excellence documented demonstration projects. Murray and Langford thought the 'demands on the industry cannot be met and so lead to an industry that cannot attract staff to deliver buildings on time, with increased costs and questionable quality.'[231] Other critics attacked the reform movement for its technocratic and managerial approach[232] and the language used.[233] By 2009 there had been little change in the industry, clients awarded projects to the lowest bidder while contractors offloaded risks and maximised profits.[234]

That a series of UK reports were required, averaging over two a decade for 50 years (many others were not included in Murray and Langford), shows how ineffective they were in developing policies to address the issues raised. The explanation for this policy ineffectiveness offered by Latham and Egan in their reports was industry culture, broadly seen as the custom and practices underlying the business model in UK construction. Latham focused on procurement and contractual relations with recommendations to change an adversarial culture, calling for more collaboration between clients, contractors, subcontractors and consultants, and more cooperative practices. He recommended 'Partnering' between clients and contractors to realise this.

Culture is clearly important, but it is also clear that culture is not malleable and does not change easily or quickly.[235] A better explanation for the lack of impact of these reports and their recommendations, and the ineffectiveness of public policy in reforming construction is required. Is the problem the policy making process, resistant to evidence and subject to ministerial whims and churn, with issues becoming politicised once they enter public debate.[236] In a technocratic system of production like construction regulatory proposals often lack a convincing evidence base, with poor integration of impact assessment and policy development processes.[237] The generic 'problem-inspired' industrial strategies developed by central policy-makers have to be interpreted by 'problem-solving' implementers responding to nuances of local context and capability.[238]

Construction is better viewed as three industries when the differences between residential building, non-residential building and engineering construction are taken into account.[239] Within the broad culture of construction they have their own permeable but distinct subcultures, based on differences in processes and products and markets. If the culture in each of the three industries is different, recommendations and policy directed at construction as a single industry are unlikely to be relevant across the three, and will thus be disregarded by many firms and clients. Clients across these industries are also different and can be generalised as households, businesses and the public sector, and their relationships with contractors varies accordingly. Another example is design, where house builders have pattern books, commercial building uses architects, and infrastructure is designed by engineers.

These structural differences between the three industries affects the way clients, contractors, designers and suppliers will interact, thus each industry has developed individual characteristics over time within the broader culture of construction that become that particular industry's subculture. The specific nature of these industry subcultures often makes recommendations and policy directed at construction as a single industry ineffective. With separate industries and separate subcultures, separate policies are required. A broad industry policy of the sort that targets construction as a single industry will be challenged by three deeply entrenched subcultures with limited, though important, similarities. Research and reports that treat construction as a single industry share this problem.

THE UK CONSTRUCTION STRATEGY

BIM mandates are important because the use of BIM unlocks the potential of digital construction and affects all the various suppliers of materials, products and services for construction of the built environment. The deeply embedded nature of the culture and processes of this production system, and the large number of small firms involved, slows technological diffusion and limits voluntary uptake of new technologies like BIM. Therefore, government and private sector clients mandating BIM is needed to pull firms along the development path from BIM level 1 to level 3. The UK is a good example. The experience of the UK after 2011 in promoting the use of BIM is an industry policy that worked, after the UK government launched an industrial strategy to improve competitiveness with a BIM mandate for public construction supported by revisions to standards. A decade later the UK was a leading user of BIM,[240] along with other early movers with BIM mandates like Singapore and Norway.

Construction industry policies in the UK moved on from the industry culture debate, although the government's objective to improve construction productivity through better procurement remained. With the launch of the Government Construction Strategy 2011-2015 and the Government Construction Strategy 2016-20 increasing the use of BIM became the target. The 2011 policy required BIM Level 2 across centrally funded construction projects by 2016, with BIM operating within the existing construction contractual framework using a legal agreement (the CIC BIM Protocol) added to professional services and construction contracts.[241] The 2016 strategy required Level 3 BIM for public projects by 2020. BIM maturity levels were defined as:

> No BIM: Information is generated manually by hand
> Level 0: Basic 2D Computer-Aided Design (CAD) use for minimal collaboration.
> Level 1: Use of 3D and 2D CAD for documentation and works information.
> Level 2: Models are shared between the project team using a common data environment.
> Level 3: Wholly integrated information model across the project, with the team working collaboratively in real-time.

The Government Construction Strategy was within the broader 2011 UK Industrial Strategy, which included Construction 2025 and targeted a 33% cost reduction in the initial costs of construction and whole life cost of built assets, 50% faster delivery from inception to completion for new build and refurbished assets, 50% lower greenhouse emissions on construction projects, and a 50% reduction in the trade gap for construction products and materials.[242] Further initiatives to support the policy were the Centre for Digital Built Britain in 2017, at the University of Cambridge, and the Construction Innovation Hub in 2018, a collaboration between the Centre for Digital Built Britain, BRE and the Manufacturing Technology Centre with £72m in government funding develop digital and manufacturing technologies in construction. The revised UK Industrial Strategy from the Department for Business, Energy and Industrial Strategy in 2017 included funding for a Construction Sector Deal. In the 2017 Industrial Strategy the government committed to Modern Methods of Construction through offsite construction for relevant departments from 2019. This was followed by the publication of Transforming Infrastructure

Performance by the Infrastructure and Projects Authority (2017, updated 2021), setting out a long term programme to improve performance and delivery. Finally, in 2018 a BIM Framework[243] based on a new ISO 19650 series of standards was released.

Ten years after the launch of the Construction Strategy progress towards BIM Level 3 remained patchy. Architects, engineers and large contractors have adopted BIM faster than services engineers, facilities managers and smaller contractors employing less than 50 people. One annual survey[244] found nearly half the 200 respondents used BIM infrequently and thought adoption was proceeding slowly, the other half used BIM often or very often. Only 6% thought progress was rapid, although 14% were using ISO 19650. Another 2021 survey by the UK BIM Alliance[245] with over 1,100 respondents found 65% were implementing BIM and used it on around half their projects and 30% were using ISO 19650. However, over half the subcontractors and cost consultants, and over 40% of project managers and facility managers, were not implementing BIM. Nevertheless, 56% of respondents thought BIM would become business as usual in 3-5 years and the other 44% thought it would take longer. Any industry strategy that approaches a technology adoption target of 100% in less than two decades has to be regarded as effective.

Compared to the limited effects of the construction reform movement's promotion of modern construction methods and OSM, which remains confined to limited markets, the BIM strategy has seen a significant increase in the use of BIM and the UK is seen as a leader in adoption[246]. The government mandate on use of BIM on public projects has been much more effective in 10 years than six decades of exhortations and recommendations to change industry culture. Recognising this, the provision of clauses covering contentious issues in construction contracts (such as intellectual property and data ownership) worked with rather than against industry practice and culture. The BIM Framework provided a roadmap for the firms and clients and the development of standards provided a toolkit. Also, local governments, universities, regulators and industry bodies were all given significant but loosely specified roles in these policies to support industry engagement.

The UK construction strategy applied to all firms involved in projects, and thus included designers, consultants and suppliers as well as contractors and subcontractors. The strategy targeted technology adoption not the 'construction industry', which is really three separate industries of residential building, non-residential building and engineering construction each with distinctive characteristics.[247] The differences in the subcultures of these separate industries accounts for the differing rates of uptake of BIM found across firms in the UK since the launch of the strategy.

Industry culture is a complex outcome of social, institutional and economic factors.[248] Because of the range and dynamic interplay of those factors it is not an appropriate target for industry policy, as the history of construction reform efforts that argued cultural change was necessary for industry improvement in the UK, documented over decades in a series of reports, clearly shows. When a new construction strategy was launched in 2011 the focus shifted from using public procurement to foster cultural change to requiring BIM on public projects, and over the next decade succeeded in increasing the use of BIM to around half of firms and the majority of public projects. Despite all the claims made for BIM changing industry culture and increasing collaboration,[249] if it were to come about it would be as a consequence not a cause of industry improvement from the new construction strategy. Recognising this, the provision of clauses covering contentious issues in construction contracts (such as intellectual property and data ownership) worked with rather than against industry practice and culture.

Compared to the limited effects of the construction reform movement's promotion of modern construction methods and OSM, which remains confined to specific markets, the BIM strategy saw a significant increase in the use of BIM. The government BIM mandate on public projects was much more effective in 10 years than the previous six decades of exhortations and recommendations to change industry culture. The BIM Framework provided a roadmap for the firms and clients and the development of standards provided a toolkit.

Another aspect of construction industry culture is that the nature of the work attracts many people with technical skills who use 'technological thinking' to find solutions to the various problems a project will encounter between inception and delivery. Technological thinking is essentially problem-solving through trial and error. Regardless of which part of construction they work in, for the vast majority of these people there is a great deal of satisfaction in doing this work well, following relevant codes of practice and meeting the required standards. Basing policies to improve industry performance and the quality of buildings on technocratic measures like ISO accreditation and BIM use levels works with industry culture.

Building Standards and Codes

The regular revision and upgrading of building codes and product standards is another policy area where governments, usually through regulatory agencies, have influenced and directed industry development. This is a more complex proposition than the use of BIM mandates because there is often no specific policy or set of policies that would qualify as an industrial strategy. Nevertheless, the use of building codes to influence industry development has a long history with some notable successes, because buildings are designed and delivered in conformance with those regulations. The building code of 1676 for the rebuilding of London after the Great Fire of 1666 classified buildings into types with specified materials and levied fees that paid for inspections. A new building code in 1844 included regulations for height, area, and occupancy of buildings.

Standards and codes establish allowable tolerances and how much variation is allowed for products and processes. They underpin quality control and are the basis of inspections to verify work being done, so a standard is a document structured around requirements for conformity and measures that certify meeting those requirements. During the late nineteenth century governments and insurers began raising the standards they set in building codes for access, light, safety, amenity and appearance, significantly improving the design and construction of buildings.[250]

The first standard was agreed in Paris for the International System of Electrical and Magnetic Units in 1881, and the International Electrotechnical Commission was established in 1906 to develop and distribute standards for the units of measurement used today. The British Standards Institution was founded in 1901, as were French and German institutes. In the US the Underwriters Laboratory was founded in 1894 by William Merrill, an electrical engineer, to provide testing of building materials for insurers, and the 1897 National Electrical Code on electrical wiring and equipment installation was the first US modern code. Insurers led the way in developing standards and methods for fireproofing the steel framed buildings that were becoming common, issuing a model building code in

1905 to reduce fire risk. Also in the US, the American Society for Testing Materials goes back to 1898 with their standard for the steel used to fabricate railway tracks. In 1902 it became the American Section of the International Association for Testing Materials, which eventually became the International Organization for Standards (ISO) in 1947. The American National Standards Institute was formed in 1918.

The ISO now has more than 22,000 different standards covering every aspect of organization management and production control. National testing and standards institutes are members of the ISO, they meet annually to review programs, and countries fund it in proportion to their trade and GDP. There is a six stage process for getting a standard published, typically based on research from the member institutes, and each standard has a guide for developing and maintaining it. Multiple standards are being combined to make them easier to manage.[251] Although agreeing new standards is a lengthy process, they are universally accepted and applied because of the rigorous scientific and engineering research they are based on. Therefore, an important element in a strategy to increase innovation in construction of the built environment is to increase funding for testing laboratories.

Building characteristics like materials, access, ventilation and fire safety are regulated by standards and codes. The International Code Council produces a series of model International Building Codes that are widely used.[252] Accreditation for standards like quality control, project management and digital twins for contractors are often required by clients. The performance of the built environment is to a large degree measured against the baselines set by standards for health and safety, energy and environmental management, and process control. When natural disasters like earthquakes, floods and hurricanes reveal shortcomings in existing standards, they lead to new standards and building code revisions.[253] The higher standards improve resilience and drive improvements in the performance of buildings and structures. This is seen when rebuilding after fires with more fire resistant buildings due to code changes, or after earthquakes with updated standards and more durable designs. Seismic code provisions first appeared in Italy and Japan in the early twentieth century, and in the US in 1927.

Building codes establish a baseline for quality and performance. They protect buildings and people from collapse, fire, wind and other extreme events.

They regulate structural integrity, electrical, plumbing and mechanical systems and safety, accessibility and energy efficiency. Codes thus underpin the work of architects, engineers, contractors and developers. Architects and engineers must ensure their building designs meet or exceed minimum code requirements Local authorities review plans before construction and issue permits. Inspectors verify the project is compliant.

Through revisions to building standards and codes innovations and new products are introduced in an incremental but typically slow process. While that reduces risk for designers and contractors, it also affects the rate of built environment product innovation and improved building performance. Revisions can be opposed or delayed, for example by residential builders worried about increased costs in a price sensitive market or by product manufacturers protecting market share. Nevertheless, a regular review and update process like the US three year cycle for building codes keeps them relevant and focused on the key issues of building quality, energy use and embodied carbon emissions from construction of the built environment.[254]

BUILT ENVIRONMENT DECARBONISATION

The role of building standards and codes in decarbonisation,[255] reducing energy use and cutting greenhouse gas (GHG) emissions is well known.[256] A carbon budget for both the construction[257] and operation[258] of the built environment is required. The UN produces an annual *Global Status Report for Buildings and Construction* that says: 'Cutting building-related emissions by improving energy efficiency is a crucial aspect of meeting net zero by 2050 climate change goals. Building energy codes provide a tool for governments to mandate the construction and maintenance of low-energy buildings.'[259]

To do this, the energy use of buildings must be monitored and managed, and buildings must be built and retrofitted to use less energy, and a global standard for determining greenhouse gas emissions for cities is under development.[260] There are many startups in building energy management. Although many countries, particularly in Africa, have not yet got compulsory energy codes, countries with codes have been moving toward electrification of building operations, particularly for heating and cooking. This is a necessary requirement to reach net zero by 2050 because residential energy use accounts for around 40 percent of total emissions.

The EU is committed to net-zero carbon emissions by 2050.[261] EU countries' national climate plans outline how a country intends to address energy efficiency, renewables and GHG emission reduction and meet EU targets. The Energy Efficiency Directive (EED), the framework for energy-efficiency policy in the EU, was established in 2012 with a 20 precent energy-efficiency target by 2020 and revised in 2018 with a 32.5 percent non-binding energy-efficiency target for 2030, with an increase to 39 percent proposed. The EED also targets government buildings, requiring renovation of 3 percent of the floor buildings owned and occupied in line to minimum energy-performance requirements.

Legislation is based on the Energy Performance of Buildings Directive (EPBD). The 2018 amendments aim for full decarbonisation of Europe's building stock by 2050 while focusing on how to modernize the existing stock. The EPBD requires Member States to develop national long-term renovation strategies, outlining how a country aims to decarbonize the building stock by 2050. To reach the '2030 climate target of reducing GHG emissions by at least 55% compared to 1990, and climate neutrality by 2050, the EU must significantly increase its rate and depth of renovation, reduce GHG emissions from buildings by 60% compared to 2015, and by 2030 increase the deep renovation rate to 3% annually, up from the current 0.2%.'[262]

There is no national energy code in the US, where state, county and city authorities all play a role in setting standards and codes. Model energy codes are developed through the International Code Council and the American Society for Heating, Refrigeration and Air-Conditioning Engineers (ASHRAE). Residential and commercial buildings typically reference a version of the International Energy Conservation Code, but California and Washington have their own codes. Codes are typically decided at local or municipal level, then adopted by the state level. New York City will phase-out fossil fuel combustion in new buildings from 2024, as will San Francisco and more than 40 other cities in the Bay Area.

More than three-dozen US cities have benchmarking policies where owners report energy data annually to local government. Some also have building labelling, which requires owners to display an energy score or ranking based on benchmarked data.[263] Building performance standards set energy or emissions targets using a range of metrics, including energy intensity, GHG emissions intensity, or third-party scoring (like an Energy

Star rating) for existing buildings. They get stricter over time, and as well as metrics and a target they include a plan of steps to be taken to reach the target. In 2022 eight US jurisdictions have implemented them. The graph below shows how improvements to the ASHRAE energy code are expected to close the gap to the 2030 target.

Figure 10. Energy efficiency and ASHRAE codes

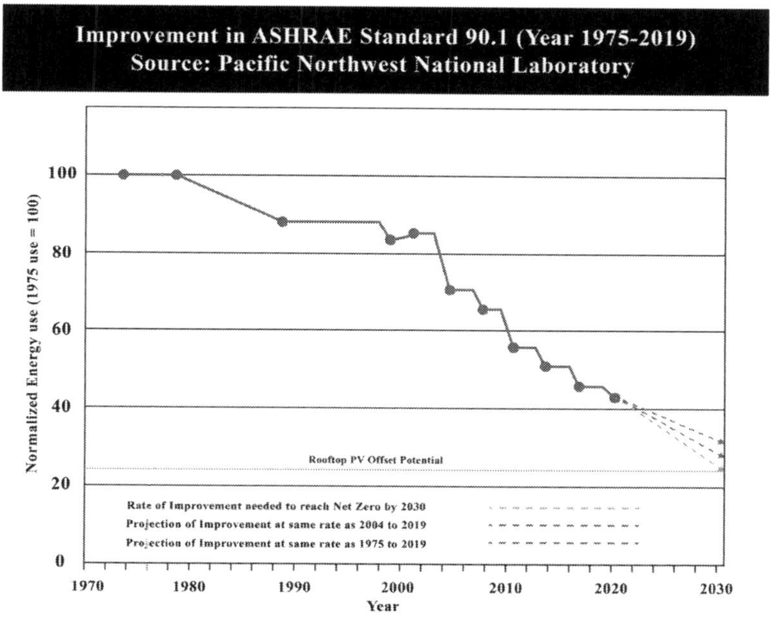

Source: Institute for Market Transformation, 2022. *Mapping US energy policy on energy efficiency in buildings*[264]

In many other countries sub-national local or regional authorities have been leading on climate change. For example, in Australia the Federal Government's 26 percent target reduction for 2030 greenhouse gas emissions is significantly lower than State Governments' 50 percent target. California is another example. It was the first US state to introduce an energy code in 1978, with the three year review and update cycle used in the US. Major updates included electric vehicle charging measures in 2015 and a rooftop solar mandate in 2020. California's 2022 building code update is considering all-electric construction, meaning buildings must use electric heating and cooking appliances, with no option to use gas.[265] New York City in 2016 required benchmarking of energy and water use and from 2020 buildings had to display their grades (from A to F) in their entrance, based on the US Energy Star system.[266]

7 Three Pathways to Future Construction

The continuity of technology is fractured by difficulties with achieving satisfactory performance in well-established methods of construction and servicing of buildings
Steven Groak[267]

ONE OF THE curious things about construction of the built environment is the perception of it as inefficient and technologically backward, yet it has been at the forefront of many scientific and technical advance for centuries. From Gothic cathedrals to railways and airport terminals, building and construction projects have bought together the best available resources to create increasingly complex structures using the best available technology. Demand for new types of structures with greatly improved capabilities in strength and span drove the development of the modern industry as it emerged during the first industrial revolution in the early nineteenth century, with the roles of engineers, architects, quantity surveyors, contractors, subcontractors and suppliers well defined by the beginning of the twentieth century.

Technologies are created through innovation, which is the application of new knowledge to develop new products and processes. Innovations, in turn, can be incremental or radical, with incremental innovation improving an existing product or process over time. Edison and Swan separately patented their designs for the incandescent electric light bulb in 1880 and, after they started working together in 1883, all subsequent improvements in the filament were incremental innovations in lighting. Radical innovation introduces a novel, new technology or idea that had not been previously available, and the invention of incandescent lighting, that produced light by running electricity through a filament, was a major advance over light made by burning oil or kerosene. Momentum is acquired as firms start using new production processes and new products are developed using the technology. As a major new technology diffuses through the economy and society, densely networked collections of industries with deep layers of specialization and complex supply chains emerge, such as electricity generation or the automobile industry.

Construction today is the outcome of such a development path. The modern industry has its roots in the beginning of industrialization in the early nineteenth century, a period of rapid, disruptive technological development not unlike the present one. Between 1850 and 1900 the building and construction industry was transformed, as an industry by the rise of large, international contractors and, over a series of major projects, by steam powered machinery and equipment and the new materials of steel, concrete and glass. On those projects new technology changed the system of production, led to the mechanization of processes, and required new methods for their organization. In the twenty-first century the effects of technological change will be heightened and quickened by the network effects associated with digitization and artificial intelligence[268].

Technological change in the form of new materials, expanding abilities and new organizational concepts is again pushing against the custom and practice that makes up the culture of a mature industry.[269] New production technologies include 3D printing (or additive manufacturing), engineered wood, integrated building design and product systems for offsite production, and temporary factories on construction sites. Advances in mechanization include both remote control and machine-control systems to automate heavy equipment such as excavators, earthmovers and cranes. Design for manufacture and assembly (DfMA) and procurement platforms integrate designers with manufacturers and manage logistics and delivery.

This chapter discusses technology and its dynamics in the fourth industrial revolution. It argues the physical characteristics of construction and the dynamics of technology lead to three distinct development paths for firms in building and construction. Firms can follow a high, low or medium technological trajectory, called the business-as-usual, upgraded-and-modified, and hybrid production paths. To an extent, these are a continuation of present trends in construction because of the constraints of building codes and regulations on one hand and the caution of firms committing to technologically driven investment on the other. The different trajectories of technological development for different firms is driven by their investment in new technologies and the rate at which they are taken up. Each path creates a context, and the technology diffusion approach discussed below indicates what can be reasonably expected over the next few decades.

Firms sorting themselves into following either a low or a high-tech path when a new technology becomes available has been seen before. In his history of the internet and broadband, economist Shane Greenstein found broadband followed three diffusion trajectories: first, broadband access diffused to households; second, broadband diffused to business, complemented by investment in advanced computing technologies; and third, different specialists offered applications. The uneven supply of networking across the US created regional variance in quantity and quality of digital infrastructure, while uneven investment created variance in quantity and quality across countries. The different patterns of application development that emerged in countries and industries was a result of the uneven distribution of digital infrastructure.[270] The adoption of broadband followed the familiar S-shaped product life cycle curve as consumers adopted first then, later in the adoption cycle, it was rising broadband speeds and more intensive content that led demand, not lower prices. This process was a repeat of earlier business transformations that followed transportation revolutions in sea shipping, railroads, and highways that also required large capital investments and had strong network effects.

Divergence between firms was reinforced by application development that targeted early adopters of the internet on the high-tech path. Over time these applications became more useful and user friendly as they spread first to followers then to mass market users. Diffusion of the internet in its first decade therefore created two investment trajectories for US firms, although by 2005 about 90 percent had adopted the basic internet.[271] Basic investment involved access to email and browsing. Advanced investment involved altering firm processes (e.g. to supply services or to receive inputs) and only 12 percent had adopted it by 2005. These investments were costly, depended on coordinating with partners, and required many complementary investments for electronic commerce,[272] Broadband use was associated with restructuring the location and scale of activity, and the use of cloud computing in US manufacturing firms predicted productivity growth among young firms and new units in established firms.[273] Frontier firms have all these characteristics.

The argument here is that, over the next few decades, the diffusion and spread of new production technologies will deeply affect how construction delivers buildings and structures. In particular, the choices available between onsite, nearsite and offsite production of many components could broaden considerably if 3D printing and digital fabrication capabilities increase and achieve the economies of scale which support offsite production. In coming decades, however, advances in automation and mechanization have the potential to significantly increase onsite production of both structural elements and components. The high-tech path is a combination of AI and digitally enabled technologies, here called 'hybrid production', which in turn becomes a new production system for delivery of buildings and structures.

Low-Tech: Business as Usual

The business-as-usual pathway is similar but smarter, relying on the slow accretive method of innovation, testing and adoption that has been followed for many decades. This is where the industry as a whole is much larger than any given project, and small individual projects reflect a consensus view on what the appropriate technological mix might be for that type of project, in that place at that time. Products are standardized and processes are flexible.

Over time this industry consensus moves to include whatever the most effective or efficient piece of technology is, again for the circumstances of the particular project and those involved. This is not a static process. The reality is that construction accepts new products from its traditional suppliers in manufacturing, plant and equipment and materials industries. The ongoing interest in offsite fabrication in all its forms indicates the industry will consider moving to a new technological platform, but the cost effectiveness and reliability of a platform has to be well proven before it becomes successful and widespread. Because there are many small and medium size standardized projects and the large and increasing stock of buildings and structures that need repair and maintenance, there will always be a market for local firms following this path. Estimates of the value of construction R&M are up to half the value of work done in new construction.

Figure 11. Pathway 1: Low tech

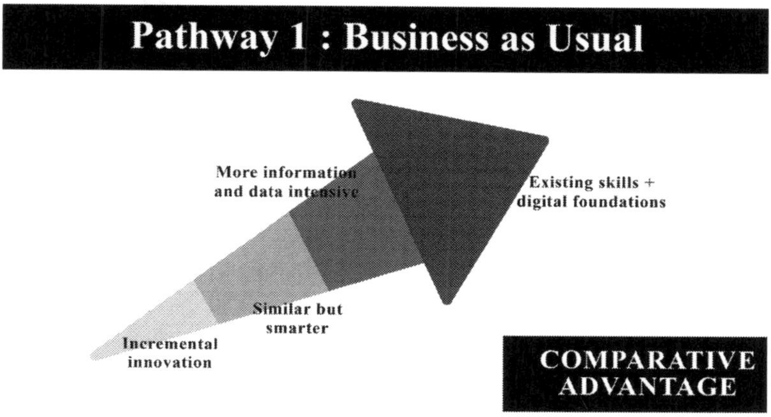

On this path firms rely on better use of resources and increased efficiency. This is comparative advantage, all firms have access to the same technology and can survive by continual, but gradual, improvement. Firms are subject to the 'ratchet effect', seen as an unintended consequence of Soviet planning.[274] If a factory met or exceeded its planned target, the target for subsequent years was increased, thus reducing effort incentives for the factory manager. In labour markets, the ratchet effect was also identified in the 1930s by sociologists studying workers subject to performance pay, who choose to restrict their output because they rationally anticipate that employers will respond to higher output by raising output requirements by cutting piecework pay or worker incentives within firms.[275]

When bidding for projects tenderers will typically not deviate far from a client's expected cost for the project[276], and all tenderers have access to the same information.[277] Because of the ratchet effect, a firm avoids revealing a significant cost advantage on one project that might jeopardise margins on future projects. Importantly, it allows for gradual improvements in productivity and efficiency, investments which are neither disruptive nor expensive to contractors, but will deliver a windfall gain to the contractor if a project comes in well under budget or schedule, which may be the result of some innovation by the contractor and will, of course, be hidden from the client and competitors as much as possible. This suggests that there might be cost-reducing innovations available to contractors, but the pressure to apply them will be affected by upfront costs, market conditions and a competitor's likelihood of using them.

This sort of low-tech future is usually missing from technology scenarios and futures studies, however a 2016 Australian analysis called *Farsight for Construction* that looked at the future of construction included it. Their scenarios describe 'four plausible futures for [the] construction industry over the coming two decades, with a focus on impacts for jobs and skills.' Each scenario consists of a description of the construction industry in the year 2036, a narrative of how the scenario came about, and a commentary on plausibility. Scenario 1's Digital Evolution is a low-tech business as usual pathway. Scenario 2's Smart Collaboration is more like a medium tech path, and Scenario 4's Rise of the Robots has elements of the high-tech path discussed below.

Figure 12. Innovation and automation

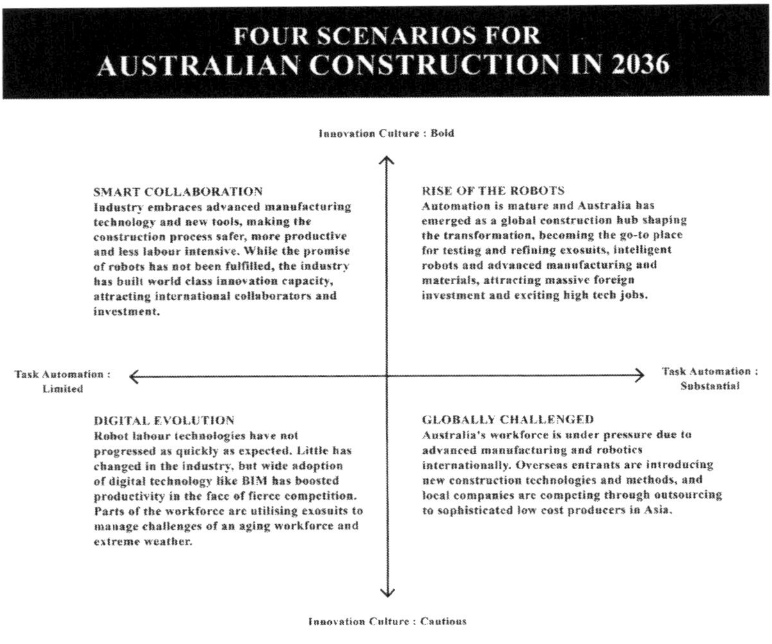

Source: Quezada et al. 2016. *Farsight for Construction: Exploratory scenarios for Queensland's construction industry to 2036.*

There is understandable skepticism across the industry about the extent of impact of new technology, the 'business as usual' approach has worked well for many small firms. Residential construction, for example, is highly decentralized, with thousands of small local firms without the capital needed for R&D and innovation, and do not implement new advances until their costs and efficiencies are proven. Innovation comes from suppliers, whose products have to reach a significant threshold for proving costs and quality. Potentially disruptive technologies such as building systems are frequently launched in market niches, but eventual scaling has proved to be difficult.

Small firms, and informal building in general, have a different trajectory of technological development to firms doing new work in the commercial contracting sector. Mainly because they are typically late adopters of new technology and follow larger firms, but also because of their continuing use of existing materials and processes. Old technologies can survive for many decades after innovations that were claimed to replace them arrived.[278] Stone, brick and wood have been used for millennia, industrialized materials like corrugated iron and concrete are ubiquitous, and a large part of construction work is maintaining and repairing the existing stock of buildings and structures. Current skills, technologies and materials will continue to be used far into the future.

Importantly, there is a class of nimble, fast growing firms that are technology leaders, some of whom are incumbents but often are not. These are 'frontier firms', or firms pushing at the technological frontier through experimentation and development.[279] Frontier firms bring with them radical new production technologies. Some of these firms are new entrants, but incumbents are also on the frontier. Examples are Trimble and Microsoft, Autodesk, Skanska embedding wireless sensors in buildings, ARUP's data collection systems and remote-controlled excavators from Caterpillar and Komatsu.

Technologies have to be adapted to become solutions for specific tasks, with their role in the workplace evolving over time to reach the levels of performance required. Only when their cost is low enough will they be adopted, but developing and engineering new technologies takes time and money. Once established, the single most important factor in technology uptake is the price/performance relationship, or the gain in productivity or other measure (time, quality, safety) the new technology delivers for a given level of investment. To successfully displace an older technology a new technology has to provide an overwhelming economic advantage to overcome the inbuilt conservatism of the existing industry, due to the investment by incumbents in the current system. This is a significant barrier because, at present, the technology price/performance trade-off is not there for the many small firms in construction because the investment required to upgrade capability is too large for the size of these firms. However, for medium and large firms the issue is not whether to invest but where and how much, if they intend to compete with frontier forms in the 'upgraded and modified' category of firms in the industry.

Medium Tech: An Upgraded and Modified Industry

The difference between this business-as-usual and the upgraded-and-modified pathway is the rate at which firms invest in and adopt new tech. There will be a widening divergence between firms that are comfortable with business as usual and familiar technology, and firms looking for ways to create or sustain market position using some combination of fourth industrial revolution technologies. Again, this might be as much about circumstances, where demand exists and the opportunity presents itself, firms would be expected to upgrade. Because of the localized nature of construction there will be large numbers of small, low productivity firms in the future. High productivity firms at the technological frontier will innovate to take advantage of technological opportunity.

If we think of the structure of the construction industry as a pyramid of different sized firms, there is a layer of tradesmen and a broad base of many small firms doing standardized work at the bottom, followed by a deep layer of medium sized firms of widely varying capability, then a small top triangle with a few large firms. Some of those large firms, and some of their major clients, are clearly on the technological frontier, and their investment in capability and capacity should deliver significant increases in efficiency and productivity, and probably scale. Some medium-size firms are also making these investments, and also have access to technologies like digital twins, drones, algorithmic optimisation, platform-based project management, robotic, VR and AR applications and so on.

In the upgraded-and-modified pathway firms have to invest considerably more in technological development to remain competitive. In the course of upgrading to these new technologies firms typically need to also make significant changes to the way they are organized and the way they organize their projects. To really leverage the investment and get an advantage from the technology, whatever it is, usually requires modification of existing business processes, and depending on how the business approaches the task these modifications could be extreme or could be at the margin. Some businesses are much better at this than others. Some of these firms might be new entrants to construction and the built environment sector, like L&G and Katerra.

The rate of adoption of technologies within firms that make up an industry is affected by a range of factors, and the technology adoption literature discusses rank effects, which are the different individual characteristics of firms such as their size, and how they affect the rate and extent of adoption of new technologies, also the effects of competitive dynamics, which is how the adoption of new technology by one company in an industry influences the adoption of technology by other companies in that industry. Due to the dynamics of a project-based industry, it is hard for contractors and consultants to spread costs incurred with R&D and innovation across projects. Consequently, the manufacturers and suppliers of building and construction products, machinery and equipment have done most of the research and innovation because they, like car companies, can standardize their products and spread the development costs over many clients.

Figure 13. Pathway 2: Medium tech

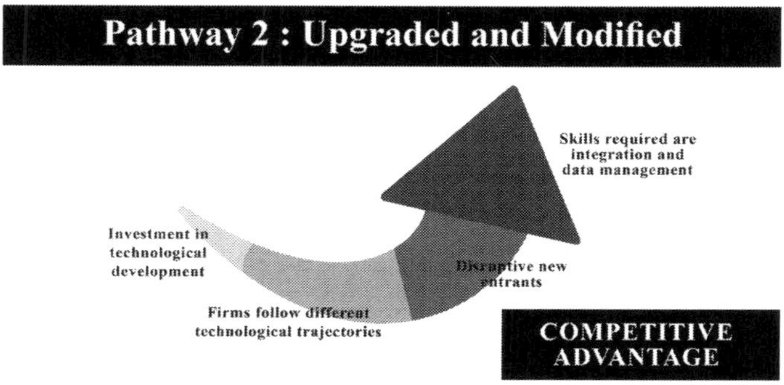

Three futures with technological context central to construction came from the World Economic Forum in their *Future Scenarios and Implications for the Construction Industry*. This scenario analysis was the second step after their *Shaping the Future of Construction* report.[280] They use infrastructure and urban development Industry (IU) to describe the built environment sector. The scenarios depict three plausible versions of the future, with implications for industry and potential winners from investment in technological capability. The WEF does not say how far into the future they are looking:

1. In *Building in a virtual world*, virtual reality touches all aspects of life, and intelligent systems and robots run the construction industry. Interconnected intelligent systems and robots run the IU, software players will gain power, and new businesses will emerge around data and services.
2. In *Factories run the world*, a corporate-dominated society uses prefabrication and modularization to create cost-efficient structures. The entire IU value chain adopts prefabrication, lean processes, and mass customization, with suppliers benefiting the most from the transition and take advantage of new business opportunities through integrated system offerings and logistics requirements.
3. In *A green reboot*, a world addressing scarce natural resources and climate change rebuilds using eco-friendly construction methods and sustainable materials. Innovative technologies, new materials and sensor-based surveillance ensure low environmental impacts, so players with deep knowledge of materials and local brownfield portfolios thrive on the new business opportunities around environmental-focused services and material recycling.

An important issue will be how the broad mass of companies in the middle of the industry, the small and medium-size contractors of every sort, cope with this tsunami of new technology as it descends over the next couple of decades. In every other industry which has become more capital intensive as technology develops, that industry has become more concentrated and the largest firms expand at the expense of mid-sized firms.[281]

This doesn't mean we end up with a few giant construction companies, but it does mean that we are likely to see a smaller number of, on average, larger firms as the gap between the larger, leading edge firms and smaller ones grows.[282] This gap can be expected to increase because the great majority of smaller firms cannot innovate as fast or as effectively as larger firms.

High Tech: Hybrid Construction

This pathway is a high-tech version of rapid and sustained advances across a broad front of key technologies deeply affecting the three dimensions of industry development: production of materials; mechanization and automation of work; and organization of projects. It is hybrid on two levels, combining human and machine intelligence[283] on one and digital fabrication with project organization on another:

- Humans collaborate with machine intelligence to design and organize projects, use robots to accomplish many tasks, or exoskeletons for physical work, and remote control automated plant and equipment.
- New production technology automates tasks and processes, new machines will become more capable and materials and machinery will become smarter as they gain embedded processors, are networked and communicate with each other. Digitally fabricated and 3D printed modular components combine with automated systems and onsite robots monitored by AI.

This is a far more speculative pathway, and the future is inherently unpredictable. The primary drivers of change on this path are AI, automation and robotics linking with digital mapping and surveying, 5D BIM, production process automation, advanced analytics, and the Internet of Things. Continued progress in molecular engineering can lead to new high performance materials and new production processes like 3D printing, and roller press printing of smart materials and fabrics. Ambitious projects will act as proving grounds for hybrid construction and their demonstration effect will ripple through the industry.

An important unknown factor is the sustainability of the recent rapid research and hardware driven gains in AI performance, however there is no reason to think performance will not continue to improve. The big unknown is how long a transition to a new technological system of production might take. While machines can replicate individual capabilities, integrating different capabilities into solutions where everything works together is another matter. Combining a range of technologies is essential for workplace automation, but solving specific problems is generally done as a series of technical challenges, and once the technical feasibility has been resolved and the technologies become commercially available it can take many years before they are adopted.

What is often missing from discussion of technological development is an appreciation of how an industry as large and diverse as construction actually adopts and implements new technology. While it's obvious that the industry as a whole is not going to be overwhelmed by some sudden mass movement to adopt some particular technology, whether it be 3D printing or BIM, the rapid pace of technological change is affecting the economic role and structure of construction. Like other industries there is potential for new entrants, new business models and great disruption. Building and construction is already using autonomous vehicles and drones, and the relevance of robotics is clear, and how those technologies might develop is becoming clearer as their use increases. The effects of AI and machine learning is less obvious, as is the spread of new technologies associated with machine intelligence.

Figure 14. Pathway 3: High tech

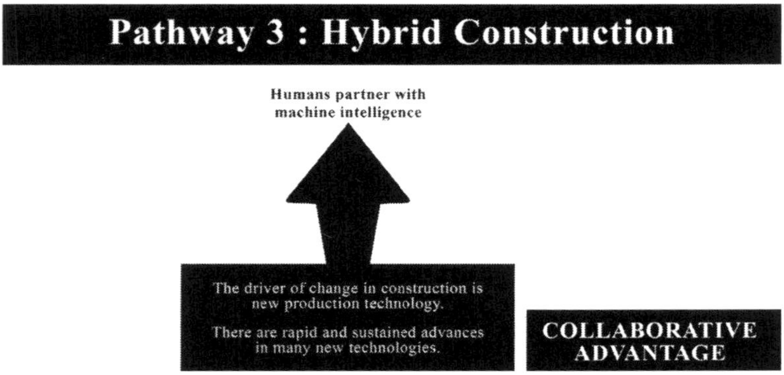

Management consultants McKinsey identified five AI-powered applications already used in other industries that could be applied to construction in 2018, in one of a series of papers on AI, automation and infrastructure as McKinsey positioned themselves for the global infrastructure boom they forecast over the next few decades.[284] The five industry applications were:

- Transportation route optimization algorithms for project planning optimization;
- Pharmaceutical outcomes prediction for constructability issues;
- Retail supply chain optimization for materials and inventory management;
- Robotics for modular or prefabrication construction and 3-D printing;
- Healthcare image recognition for risk and safety management.

Each of these applications had a short discussion with examples of crossover potential. Using AI for optimization is obvious, McKinsey looked at machine learning algorithms that are relevant to contractors and briefly assessed their potential. They suggest algorithms will be useful for: refining quality control and claims management; increasing talent retention and development; boosting project monitoring and risk management; and constant design optimization.

How firms use technology, in the way it is adopted, adapted and applied, varies widely within construction. However, current materials and methods have been used and refined for over 100 years. With the development of plate glass, steel reinforced concrete and standardized components, construction technology is an established, mature system of production. Mature technologies become conservative as they accumulate capital and skills, and this investment in the existing system gives it great inertia until some disruptive change emerges. The culture of an industry develops over many years and will resist change that threatens the roles of incumbents. However, industry is only one of many factors and determining how and why a new technology spreads through the economy and society. Studies of historical cases like steam power, tractors, electricity, TV and phones have given us good examples of technology diffusion, which typically takes decades, and its dynamics as new technologies become more competitive.

A sign of the maturing of digital construction is the uptake of ISO 19650 standards and the widespread recognition of the reality of a digital future, broadly based on BIM and digital twins. Will construction industry development over the next decades absorb the impacts of this new technology without significant effects on industry structure or dynamics? Given the entanglement of economic, social, political, and legal factors in the construction technological system this might be the case, however there are good reasons to think this may be wrong. Machine learning, AI, automation and robotics are an interconnected set of technologies that are evolving quickly, enabled by expanding connectivity and the massively scaleable hardware available today.

This hybrid construction path based on AI does not have any current examples, although key elements such as robotic equipment, drones and monitoring systems are in use. The characteristics of this future industry might best be seen in what could produce, which would be smart and responsive buildings and structures, made of high-performance smart materials which know their location, purpose and condition, run by operating systems that constantly monitor and control a building's internal environment and systems, and with an energy efficient, self-repairing external skin. To deliver and operate such a building will be a data-intensive task, as its digital twin takes shape decisions on what and where to produce the components will be made, and what machinery and equipment will be needed to install them. The project would be delivered through a massively integrated platform linking procurement, design, manufacturing and logistics entirely underpinned by digital data and AI assistants at each step. Although this is beyond the current technological frontier, many of the foundational elements are here today.

Innovation and Industry

A period of technology-driven restructuring of construction may be about to start, similar to the second half of the 1800s when the new materials of glass, steel and reinforced concrete arrived, which led to new methods of production, organization and management. There are many implications of such a restructuring. Some firms are rethinking their processes in response to developments in AI, robotics and automation as capabilities improve quickly and the range of new products using these technologies expands. The majority of firms, however, are focused on incremental improvements in current technology. Meanwhile, firms at the technology frontier are pushing the boundaries of what is possible, exploring, experimenting, and inventing. Many experiments will be costly, and firms will fail, like Katerra in 2021.

A technological trajectory where it takes two to three decades for a technology to move from the experimental periphery to widespread adoption and use is very common. The internet had been around for over 20 years before Netscape made it widely accessible in 1994 by using graphics. At that time, globally, there were about 600 websites and a couple of million connected computers. Amazon and eBay launched in 1995. There are many examples of the way technology proceeds a step at a time as the necessary system components come together and get improved. The question this discussion addresses is 'What systems around today might be the foundations of construction tomorrow?' Automation, robotics, AI, 3D printing and digital fabrication are all at the stage where they can make the transition from experimental technologies to wider use and diffusion through industry.

Over the next few decades the uneven diffusion of these new GPTs through industry and the economy will lead to three distinct development paths for firms in construction and related built environment industries. With new GPTs firms can follow one of high, medium or low technological trajectories, i.e. the business-as-usual, upgraded-and-modified, and hybrid production. What differentiates the low and medium tech pathways is investment in technological capability and technological diffusion, the rate at which these new technologies are taken up across different firms, which in turn leads to different trajectories of technological development for firms within those three pathways.[285] Every project creates a market for particular firms, and the technology diffusion approach allows some analysis of technological developments that will affect those markets and firms.[286]

The third pathway is a high-tech version of rapid and sustained advances across a broad front of technologies that combine human and machine intelligence with digitized production and project management technologies. The two primary drivers of transformational change are rapid diffusion of digitization in procurement, design and production, and the reorganization of project delivery they allow. Creation of project data sets and digital twins will enable automation and robotics to transform the organization and delivery of projects, and new materials and production processes will allow new types of buildings and greatly improve their performance. Historically, a similar transformation occurred in the second half of the nineteenth century with steam power and mechanization, creating a new system of production based on new materials and machinery. In the twenty-first century it will be AI, 3D printing and digital fabrication.

Conclusion

If we want things to stay as they are, things will have to change.
Giuseppe di Lampedusa[287]

THE SINGLE MOST important factor in technology uptake is the price/performance relationship, or the gain in productivity or other measure (time, quality, safety, choice) the new technology delivers for a given level of investment. To successfully displace an older technology a new technology has to provide an overwhelming economic advantage to overcome the inbuilt conservatism of an existing system of production, due to the investment by incumbents in the current system. An optimistic view of diffusion has continuing development of applications followed by more diffusion to more users, with a feedback loop leading to a widening set of applications. A pessimistic view emphasises the barriers to change and the difficulty of reaching viable economies of scale, shortages of skills and lack of training.

Examples of invention and innovation have been given, and their role in models of industry life cycle discussed. The history of incremental innovation in concrete was used to emphasise the importance of the relationship between construction materials, industry organization and site processes. Between 1800 and 1900 there were a series of technological shocks as methods of industrialized building with glass, steel and reinforced concrete were refined. With industrialized production, prices of manufactured products declined as economies of scale and scope were realised, and over time those cheaper prices allowed new technologies to spread and find new uses. Often a combination of two new innovations was more significant than the individual technologies.

It was argued a period of rapid development of construction and related industries may be about to start, driven by the new general purpose technologies of AI and digital fabrication using 3D printers. The only comparable period is the second half of the 1800s, when steam power and the new materials of glass, steel and reinforced concrete arrived, bringing with them new business models and new entrants. It was suggested the dimensions of industry development during the nineteenth century provide a framework for thinking about construction of the built environment in the twenty-first century. These were industrialization of production, mechanization of work, and organization of projects.

There are many issues affecting the built environment, many of which are wicked problems of great complexity that range widely across industries, institutions and regulatory systems. All costs the complex institutional and policy environment entail are crystalised at the moment a contract is signed for a new building or construction project, as part of a total cost that typically includes finance and land, or access to it. The remaining share of the project cost is design and delivery, so that is what built environment industries can affect. On the supply side the issues are about efficiency, productivity and production costs.

There are many reasons for the apparently persistently poor performance of construction. As explained by McKinsey, the industry is extensively regulated and highly dependent on public-sector demand. Informality and sometimes corruption distort the market. Construction is highly fragmented, contracts have mismatches in risk allocations and rewards, inexperienced owners and buyers find it hard to navigate an opaque marketplace. There can be poor project management and execution due to insufficient skills, inadequate design processes, and underinvestment in skills development, R&D, and innovation.[288] The increasing cost of major projects and lack of productivity growth in construction has been an issue for governments and major clients for the six decades since productivity statistics first became available in the 1960s.

This book starts from a different position, that the cost of buildings and structures is largely determined by market conditions, and they are delivered as efficiently as possible, given technology and circumstances, by a complex system of production that is well over 100 years into its life cycle. Clients are unwilling to pay for innovation and R&D because the benefits to an individual project are small but both cost and risk increase. Fragmentation is inevitable in a geographically dispersed project-based industry with volatile demand, but many parts of the supply chain are oligopolies. Much of the value added is found offsite, in design and manufacturing, not in the onsite work of contractors and subcontractors. Critics can overlook these facts. While the current system of production may be inelegant and inefficient it is flexible and resilient, delivering projects of all types under all conditions.

Despite the extent of technological change expected over the next few decades, some new industry will not appear to undertake construction of the built environment. However, that does not mean it will not be affected by creative destruction with future, but foreseeable, developments in AI, automation and robotics. In the three technology pathways outlined here construction today more or less fits into the business-as-usual approach of path one, and the global industry rather looks like it's following the upgraded and modified path in path two. These two technological trajectories cover the likely outcomes of near future developments, and they are both based on well-established fundamental characteristics and trends that we observe today. Why then have the third high tech pathway? The sort of advanced buildings and structures it envisages will not be technically feasible for some time, it could take several decades before the experimental work being done today becomes the standard technology of tomorrow.

Nevertheless, this experimental work is the basis of the future industry. For example, there is a lot of work being done in labs around the world on molecular engineering of materials and new forms of digitized production processes. Commercial applications of 3D concrete printing are available, and there is a well-established maker movement using digital fabrication laboratories. Sensors are being placed in structures to monitor their condition, scanning is replacing visual inspections for compliance, cracks and fatigue, and remote control and sensing happening. Today's existing scientific and technological base will drive the development of the industry of tomorrow (which is why government investment in basic R&D is so important).

Two decades into the twenty-first century, a cascade of new technology is fundamentally affecting construction firms, processes and products. Attention gets focused on new tech that is currently being adopted, and of course marketed. Typically, this is not really 'new' but a commercial product that is the result of years or decades of development. Examples are BIM, 3D printing, and cloud-based project management which, in all their many and various forms, are now spreading through construction after decades of development. While not understating the effects these will have, they are just the first stages in a series of technological waves expected in the near future.

From economic history, we know major new technologies take time to diffuse through the economy because they require parallel changes in forms of organization, methods of production and patterns of consumption. These are not decisions firms and households make quickly or easily, due to the investment in upgrading machinery and equipment usually needed. New technologies are 'embodied' in this new physical capital, in the way a twenty year old car incorporates the technology of two decades ago when it was made. Studies of the diffusion of new technologies like electricity, tractors and the internal combustion engine[289] find it typically takes fifteen to thirty years for a new technology to reach 90 percent or more of its potential market.

As is so often the case with technological transitions, it more likely to be new combinations of technologies and innovations that transform construction, not some breakthrough technology that leads to some different, new industry we don't already have. Combining robotic and automated machinery with 3D printing of standardized parts opens up many possibilities. Until now 3D printing has not been economic for mass production, however designing an automated onsite production process that includes the machines and equipment need to move and install the parts produced by printers is possible, moving beyond design for manufacture and assembly to an integrated production and assembly system.

Another example is digital twins. By combining BIM systems with specialist systems for surveying, design and engineering they become intelligent systems that interact with users to actively support their work through databases and access to information. This is what generative design systems are doing, not replacing people or skills but augmenting them. Digital twins can be combined with virtual reality (VR) and augmented reality (AR) systems to create a holographic virtual project that contains every detail of a building, and that information can be shared through a project management platform with project participants. The data a virtual project produces and the information flow it supports could well be transformational, and will certainly challenge existing business models. At this point the expectation is that VR will be used more on the design side by architects, planners and engineers, while AR will have a larger footprint on construction sites, although some firms have started looking at using VR in areas like safety and training.

BIM has taken the two to three decades of development to get to widespread use typical of new technology, from its origins in 1970s CAD programs. Governments and private sector clients wanting to promote BIM have successfully used mandates to increase uptake. While not a new technology, BIM is the required enabler of further developments, a necessary foundation for the transition to the system of production and delivery of projects in the digital age. Although linking BIM to manufacturing or the sharing of a single record of a project is a significant change in construction, these are not in themselves new either. In other industries techniques like design for manufacture, lean production and digital documentation are well-established, with the adoption lag (the difference in the diffusion rate) for construction largely due to its regulation, fragmentation, project-based structure and conservative clients. The role insurance companies play in building standards and codes and banks in approving finance also has to be taken into account. Because they are focused on risk minimisation they are cautious when it comes to innovation and new construction technologies.

How firms use technology, the way it is adopted, adapted and applied, varies widely within construction. This is both a significant driver of incremental change and a major inhibitor of technology diffusion. After 100 years the modern system of production based on steel, concrete and glass is fully developed, and as a mature system with established standards and codes it is also conservative. Mature systems accumulate capital and skills, and this investment in the existing system gives it great inertia until some disruptive change emerges. How and why a new technology spreads through the economy and society are determined by many factors, but historical cases like steam power, tractors, electricity, TV and phones have given us good examples of technology diffusion and its dynamics. Although it may be a drawn-out transition to a digital production system, the use of cloud based digital twins could reach 90 percent of firms in the next decade, following the path of email and browsing in the first decade of the internet.

Finally, the role of clients in creative destruction is fundamental. Their demands for buildings with improved physical and environmental performance, or new types like spaceports and automated factories, drive industry forward. From the first industrial revolution's docks, factories and warehouses to the third's airports, office towers and malls, new industries have required new buildings. Now new industries require the data centres, clean rooms and solar and wind farms powering the fourth. The most ambitious of those buildings have always been at the limits of the available technology and used advanced engineering in their design as the functions of the built environment evolve over time. These changing requirements are the demand-pull side of innovation and new technology in construction of the built environment.

Figure 15. Time and the built environment

FUNCTION	PAST	PRESENT	FUTURE
Accommodation : Where we stay			
Workplace : Where we make things			
Services : Where we get things			
Logistics : How things get to us			
Infrastructure : How it works			

Note: Maintenance and repair of the built environment accounts for a significant share of construction work, estimates range from a third to a half of the value of new work. Annual additions to the existing building stock are a small percentage of the total number of buildings.

Diffusion and Disruption

Technological developments are combining intelligent machines with engineered materials, deep learning capabilities, human augmentation and new organizational concepts, and are pushing against established custom and practice in a mature technological system. Because the system is mature the effect of new technology and the changes it brings could spread slowly across the industry as a whole, and unevenly because of the many small and medium size firms. While this was case with twentieth century GPTs like electricity, a period of disruptive change in the construction industry occurred during the second half of the nineteenth century, and a new system of production eventually resulted in a new form of industry organization led by contractors instead of architects and engineers.

That disruptive transition took several decades, as industrial materials replaced craft ones and site work was mechanized and reorganized. Then, over the twentieth century contractors evolved into project managers and the traditional trades became subcontractors.

At this point how quickly and how far AI, 3D printing and digital fabrication will spread through construction over the next decade is unknowable. There are many technical, regulatory and institutional hurdles to be overcome, so diffusion might be slow for many reasons. On the other hand, if big data, analytics and AI can improve project processes and digital fabrication increases onsite productivity, they could be adopted quickly by companies that have the systems and resources to take advantage of them. AI in construction is already in use for generative design, drone monitoring and managing equipment maintenance.

A starting date for AI in its current form might be IBM's Watson in 2011 or DeepMind's AlphaGo in 2016, so there is some way to go and further development needed before AI becomes widely dispersed if past experience with other powerful GPTs is a guide. Digital fabrication, however, is about to start its third decade of development with an already substantial footprint. Architecture and engineering schools have robotics labs and the supply chain of equipment manufacturers and materials suppliers is in place. International standards like ISO 19650 have been released. Several prototypes of 3D printed concrete houses have been completed and a range of concrete printers are commercially available. If digital fabrication delivers the economies of scale expected, it will affect all participants in the construction of the built environment.

With a technological trajectory for industry based on AI, 3D printing and digital production technologies, the view taken here is that there will be a transition period of perhaps a decade, possibly two, as construction adopts them. As that happens the organization and structure of the industry would change, from one centred on PMs and contractors to one based on integrators that combine site preparation with a mixture of onsite or nearsite production using 3D printing in modular fabs and mass-produced items.

AI as a new GPT may be as disruptive to construction as steam power was in the nineteenth century, and lead to a similar restructuring of the industry. Neither electricity nor computing had a significant effect on the organization of construction, because the evolution from contractors and trades to PMs and subcontractors was not driven by those technologies. However, the change from master builders and crafts to contractors and trades was a break from the past, and the result of industrialization and mechanization.

Broadly, there are three types of machines: human operated, some like aircraft with autopilots already a partnership with machines; remotely operated ones like drones; and autonomous machines like Mars rovers. It is not so much these machines and robots as the emerging combinations of humans and intelligent machines that is changing the nature of work. An example of a new onsite technology is automated drone monitoring with laser scanners, combing BIM and location data, as do smart helmets. Linking smart helmets with intelligent machines may address the characteristic changeability of construction sites, which is challenging for automated and robotic systems, where a human supervisor operating a team of robots or several pieces of equipment, some with limited autonomy, might work better. A worker with a smart helmet could monitor these machines both on the project site and in the digital model.

Some caution is required, however. In the early 1980s psychologist Lisanne Bainbridge pointed out the 'ironies of automation': the loss of physical skills; the role of humans changing from active control to passive observation; and automation does not make people redundant.[290] More recently, in 2022, social researcher Luke Munn described automation as 'a fable that rests on a set of triple fictions' or myths: the myth of full autonomy claims machines will take over production, but technical solutions are piecemeal; the myth of universal automation sees technologies sweeping the globe, but ignores the local social, cultural and geographical forces that shape technology use; and the myth of automating everyone ignores the reality that automation's fallout will be highly uneven.[291] These are important issues and there are many reasons for the persistence of old technologies and the slow diffusion of the new. Creative destruction works over many decades not a few years.

In industries with controlled environments like factories and production lines, automation can reduce and potentially eliminate labour. Mass production of standardized products justifies the capital investment in plant required and the market demand for products is well known and stable, unlike the highly variable demand for buildings, which rises and falls with the business cycle. The procurement and contracting system typically used in construction is for specific projects and is focused on cost and risk management, not innovation and R&D. Contractors look for process improvements that increase onsite efficiency, and rely on manufacturers and suppliers for product innovations.

It should be clear that the role of the technologies discussed here will be to augment human labour in construction of the built environment, not replace it, and over the next decade education and training pathways and industry policies with incentives for labour-friendly technology will be needed.[292] Generative design software does not replace architects or engineers. Optimization of logistics or maintenance by AI does not replace mechanics. Onsite construction is a project-based activity using standardized components to deliver a specific building or structure in a specific location. The nature of a construction site means automated machinery and equipment will have to be constantly monitored and managed by people, with many of their current skills still relevant but applied in a different way. Dangerous tasks should be done with remote controlled equipment.

Nevertheless, in the various forms that BIM, digital twins, AI, 3D printing, digital fabrication and procurement platforms take on their way to the construction site, they will become central to many of the tasks and activities involved. In this, construction may no different from other industries affected by fourth industrial revolution technologies, however the path taken will be distinct and different from the path taken in other industries. This path dependence varies not just from industry to industry, but from firm to firm as well. Because construction of the built environment as a system of production is so wide and deep, this will affect a large number of people, and through them the wider economy and society. The future is not determined, although technological change and creative destruction continue to reshape and restructure industry and the economy, decisions made today create the future.

Appendix: The Built Environment Sector and Construction Statistics

We are surrounded both by the built environment and by the continual remaking and creation of its fabric Building and construction is around seven or eight percent of gross domestic product (GDP) in industrial economies, and twice that or more in developing economies. Significantly, that onsite construction brings together an extensive network of suppliers in the production of the built environment that contributes as much again to GDP. There is a complex network of industries responsible for the organization of the design, delivery and maintenance of buildings and structures.

Onsite work links suppliers of materials, machinery and equipment, products and components, the inputs required to deliver the buildings and structures that make up the built environment. Consultants provide design, engineering, cost planning and project management services. There are also inputs from marketing, urban planning, transport, finance and legal services. Once produced, buildings and structures then need to be managed and maintained over their life cycle, work done by another group of related industries. The built environment needs infrastructure and services like water and waste disposal, provided by yet more industries. This can be thought of as the difference between an industrial cluster or sector, made up of contractors and sub-contractors supported by plant and equipment suppliers, consultants, manufacturers, distributors and others, and the onsite activity that is measured as 'construction work'.

These two views have been called the broad and narrow construction industry. The narrow industry is onsite work of contractors and subcontractors and the broad industry the value chain of materials, products and services.[293] Many industries are structured around such value chains and production networks, and when enough firms share sufficient characteristics they are often described as an industrial cluster or sector.

Construction of the Built Environment as an Industrial Sector

An industrial cluster brings together a group related firms and was originally applied to specific locations like the wine industry in California's Napa Valley or Bordeaux in France. Over time, the concept broadened as different types of clusters were identified, such as creative industry hubs or knowledge centres.[294] Two types of clusters are:

Geographical – industries using the same resources in a specific location

- Movies – Hollywood US, Bollywood India;
- IT – Silicon Valley CA, Silicon Alley NY, Silicon Glen Scotland, Bangalore India;
- Leather goods, spectacles and glasses – Milan Italy;
- Health – Boston US, Oxford England, Chennai India;
- Electronics – Guadalajara Mexico, Cordoba Argentina, Guangdong China;
- Finance – London England, New York US, Geneva Switzerland; and

Vertical – a hub and spoke value chain from suppliers to end products

- Automotive – Detroit US, Dusseldorf Germany, Turin Italy, Curitiba Brazil;
- Aerospace – Toulouse France (Airbus), Seattle US (Boeing);
- Smart phones – Guangdong China (Apple), Hanoi Vietnam (Samsung).

Some industries do not have central locations like the clusters in IT, wine and finance, or don't have major hubs where production is concentrated like automobiles and aerospace. These industries are built around decentralised production, distribution and delivery networks that make their products widely available to clients and customers. Examples are:

- Pharmaceuticals – a globally distributed industry, with countries combining some form of domestic production and imported supplies;

- Shipbuilding – brings many suppliers together in a few locations;
- Electricity generation – brings many suppliers together in many locations;

Building and construction is the world's most ubiquitous industry,[295] sharing the most widely used materials of wood, clay, glass, steel and concrete. Is this really a cluster or is it too diverse? A wide range of firms from a broad collection of industries are involved in designing, producing and maintaining the built environment. Where to draw the boundaries of the wider industry is an open question, as a diverse range of firms, professional institutions, government regulators and authorities all contribute to the creation and maintenance of the built environment.

Construction economists have advocated looking at the sector that produces the built environment in broad and integrative terms.[296] For construction, this is the difference between an industrial sector made up of contractors, sub-contractors, equipment suppliers, professional services, materials, manufacturers, and equipment suppliers and the onsite work done by contractors and sub-contractors. George Ofori wrote 'a broad definition of construction is proposed….as that sector of the economy which plans, designs, constructs, alters, maintains, repairs and eventually demolishes buildings of all kinds, civil engineering works, mechanical and electrical engineering structures and other similar works.'[297]

Figure 16. Industry inputs and outputs

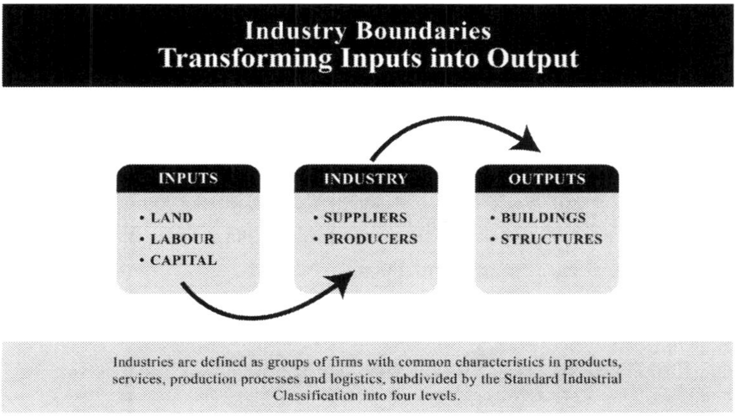

A dense network of many different firms and participants organised in a system of production is called an industrial sector, because it is too diverse and distributed to be a cluster in the conventional geographic version. There is no definition of an industrial sector, beyond a broad collection of firms with one or more common characteristics, like the 'non-profit sector', 'manufacturing' or 'the business sector'. There are also sectors based around a definable market, two examples being:

- Defence - there is no defence 'industry' because suppliers come from many different industries like IT, aerospace and shipbuilding, but as a sector share resources and clients; and
- Tourism - which brings together the contributions of industries like accommodation, tour operators and entertainment. Australia has an annual Tourism Satellite Account produced each year jointly funded by industry.[298]

Tourism is not an industry in the conventional sense and is defined in the tourism satellite account as an aggregation of other industries such as transport, accommodation, hospitality, retail trade, entertainment and education. A satellite account reclassifies different industries into a single sector and is used to provide more detail on sectors that have contributions from a number of industries. The most common satellite account is for tourism (often jointly funded by industry and users), but they have been produced for a wide range of other industries such as health, the environment, R&D, information technology, infrastructure, non-profit institutions, human capital and households.

Parts of the economy that involve many different contributors and participants are an industrial or economic sector. If the built environment encompasses the entirety of the human built world, then the collection of industries responsible for designing, producing, managing and maintaining the buildings and structures that humans build can be called the Built Environment Sector (BES). The collective significance of these industries is obscured by their diversity, ranging from architecture to waste disposal, and their geographic distribution. However, by using *Standard Industrial Classification* (SIC) data from national statistical agencies a profile can be built of the broad industry by including industries involved in construction of the built environment. Research based on SIC data has found that the BES accounts for 15 to 20 percent of GDP in OECD countries.[299]

THE INDUSTRY CLASSIFICATION SYSTEM

In the United Nations *International Standard Industrial Classification* (ISIC) economic activities are subdivided into a four-level structure. Activities are first divided into 'sections', which are alphabetically coded. These sections divide productive activities into broad groupings such as 'Agriculture, forestry and fishing' (A), 'Manufacturing' (C) and 'Information and communication' (J). The classification is then organized into numerically coded categories, which are two-digit divisions, three-digit groups, and four-digit classes. In the current ISIC system Construction is a section made up of three divisions. The Manufacturing sector has the most detail with 23 divisions, Information and communication has five divisions. Construction is Section F in ISIC and has three divisions, the Construction of buildings (division 41) which includes both residential and non-residential building, the Construction of civil engineering works (division 42), and Specialized construction activities or special trades (division 43). Alterations and additions to buildings and the repair of buildings and engineering works are intended to be included, but measurement of these activities is limited.

The first *Standard Industrial Classification of All Economic* Activities was established in the US in 1937, with the United Nations *International Standard Industrial Classification* following in 1948. This internationally accepted standard had its most recent revision in 2008:

> This fourth revision of ISIC enhances the relevance of the classification by better reflecting the current structure of the world economy, recognizing new industries that have emerged over the past 20 years and facilitating international comparison through increased comparability with existing regional classifications.[300]

Table 11. The International Standard Industrial Classification

Section	Industry	Divisions
A	Agriculture, forestry and fishing	1-3
B	Mining and quarrying	5-9
C	Manufacturing	10-33
D	Electricity, gas, steam and air conditioning supply	35
E	Water supply; sewerage, waste management and remediation	36-39
F	Construction	41-43
G	Wholesale and retail trade; repair of motor vehicles and motorcycles	45-47
H	Transportation and storage	49-53
I	Accommodation and food service activities	55-56
J	Information and communication	58-63
K	Financial and insurance activities	64-66
L	Real estate activities	68
M	Professional, scientific and technical activities	69-75
N	Administrative and support service activities	77-82
O	Public administration and defence; compulsory social security	84
P	Education	85
Q	Human health and social work activities	86-88
R	Arts, entertainment and recreation	90-93

S	Other service activities	94-96
T	Activities of households as employers and for own use	97-98
U	Activities of extraterritorial organizations and bodies	99

Different versions have developed as countries adapt the ISIC to their economic structure. Three versions are ANZSIC; the European Union (EU) NACE and United Kingdom's (UK) *Standard Industrial Classification of Economic Activities*, and the *North American Industry Classification System* (NAICS) used by the US and Statistics Canada. The three are similar to ISIC, which provides the bones of their systems, and the three systems all have separate classifications for residential and non-residential building engineering construction and specialized trades. There is an important point of difference because NACE and the UK SIC have Development of building projects as part of the building industry. Neither ISIC nor the other two systems have this subdivision.

The ISIC division Specialized construction activities covers the trades. These are divided into three classes: Demolition and site preparation; Electrical, plumbing and other installation activities; and Building completion and finishing. At this three-digit level of classes there are many differences between ISIC and each of the three SIC systems, reflecting variations in local conditions ('Siding contractors' in Canada) or industry practice ('Hire of construction machinery with operator' in Australia or 'Scaffold erection' in the UK). The NAICS has four groups: Foundation, Structure, and Building Exterior Contractors; Building Equipment Contractors; Building Finishing Contractors; and Other Specialty Trade Contractors. The ISIC explains the classifications:

> This division includes specialized construction activities (special trades), i.e. the construction of parts of buildings and civil engineering works without responsibility for the entire project. These activities are usually specialized in one aspect common to different structures, requiring specialized skills or equipment, such as pile driving, foundation work, carcass work, concrete work, brick laying, stone setting, scaffolding, roof covering, etc. The erection

of steel structures is included, provided that the parts are not produced by the same unit. Specialized construction activities are mostly carried out under subcontract, but especially in repair construction it is done directly for the owner of the property. Also included are building finishing and building completion activities. Included is the installation of all kind of utilities that make the construction function as such. These activities are usually performed at the site of the construction, although parts of the job may be carried out in a special shop. Included are activities such as plumbing, installation of heating and air-conditioning systems, antennas, alarm systems and other electrical work, sprinkler systems, elevators and escalators, etc. Also included are insulation work (water, heat, sound), sheet metal work, commercial refrigerating work, the installation of illumination and signalling systems for roads, railways, airports, harbours, etc. Also included is the repair of the same type as the above-mentioned activities.

Building completion activities encompass activities that contribute to the completion or finishing of a construction such as glazing, plastering, painting, floor and wall tiling or covering with other materials like parquet, carpets, wallpaper, etc., floor sanding, finish carpentry, acoustical work, cleaning of the exterior, etc. Also included is the repair of the same type as the above-mentioned activities.[301]

The latest versions of ISIC and these SICs are the most recent in a series of revisions that have been made over the decades, to allow the statistics published better reflect the changing structure of production in a modern economy. As industries have developed and evolved the classifications have been expanded and several sections divided, with their divisions becoming new sections. The revisions to ANZSIC shown in the table are representative of the changes that have been made, with the growing importance of service industries in post-industrial economies. In the 1994 ANZSIC revision three sections were divided into two: Wholesale and retail trade; Finance, property and business services; Community Services. One section was divided into three as Recreation, personal, other services became: Accommodation, cafes and restaurants; Cultural and recreational services; and Personal and other services.

The 2006 revision then divided two more sections: Government administration and defence became two sections; and Property and business services, which was split into Rental, hiring and real estate services and Professional, scientific and technical services.

Table 12. Revising ISIC: the ANZSIC example

ASIC 1983	ANZSIC 1994	ANZSIC 2006
Agriculture, forestry, fishing	Agriculture, forestry, fishing	Agriculture, forestry and fishing
Mining	Mining	Mining
Manufacturing	Manufacturing	Manufacturing
Electricity, gas and water	Electricity, gas and water	Electricity, gas and water supply
Construction	Construction	Construction
Wholesale and retail trade	Wholesale trade Retail trade	Wholesale trade Retail trade
Transport and storage	Transport and storage	Transport, postal and warehousing
Communication	Communication services	Information & telecommunications
Finance, property and business services	Finance and insurance Property and business services	Financial and insurance services Rental, hiring, real estate services Professional, scientific and technical

Public administration and defence	Government admin. & defence	Administrative and support services Public administration and safety
Community Services	Education Health and community services	Education and training Health care and social assistance
Recreation, personal, other services	Accommodation & restaurant Cultural & recreational Personal and other services	Accommodation and food services Arts and recreation services Other services

The way we see and understand an industry starts with the data provided by national statistics agencies. How data on Construction is collected and how the categories within Construction are defined is clearly important. The former determines the quality and the latter the credibility of the statistics produced. Construction economists Jim Meikle and Stephen Gruneberg concluded that:

> The current ISIC breakdown of construction activity is not particularly helpful to any user groups. It requires distinctions to be made among residential, non-residential and civil engineering work. It does not distinguish between construction investment (new work and improvements) or construction consumption (repair and maintenance) or among publicly sponsored, privately sponsored and mixed-funded work. Detailed breakdowns of construction activity could also address the different providers of construction output: construction contractors, the informal sector, households and so forth.[302]

Economic Role of the Australian Built Environment Sector

The idea that the construction industry is only one part of the creation and maintenance of the built environment has been developed by researchers into a method for measuring the broad construction industry. Using SIC data a profile can be built of the broad industry by including industries involved in construction of the built environment.[303] Construction economists using this method found onsite work by contractors and subcontractors accounted for half the total output and employment of the broad industry.

To measure the contribution to GDP and employment of the built environment sector BES, data on the industries included in the BES needs to be available at the required level of detail. Industry level data is available from national statistical agencies, although the frequency and formats of its publication vary widely. There is also leakage around the boundaries of industry statistics: some glass is used in mirrors, some in car windscreens; textiles are used in buildings; architects design furniture; engineers repair machines as well as structures, and so on. The SIC data does not provide that level of detail.

For Australia there is an annual series of industry data published since 2007. The Australian BES combines data for sixteen industries included in *Australian Industry* (ABS 8155) that together form one of the largest and most important industrial sectors in the economy. The analysis is based on Industry employment (number of people) and Industry value added (IVA), the estimate of an industry's output and its contribution to gross domestic product (GDP) from the difference between total income and total expenses. IVA is given in current dollars in *Australian Industry*. The data is presented at varying levels for SIC industry divisions, subdivisions and classes. The most recent issue is for 2018-19.

Table 13. Industries included in the Australian Built Environment Sector

Supply industries	Demand industries	Maintenance industries
Non-metallic mining and quarrying	Residential property	Water, sewerage and drainage
Building construction	Non-residential property	Waste collection, and disposal
Heavy and civil engineering	Real estate services	Building and industrial cleaning
Construction services		Building pest control services
Architectural services		Gardening services
Surveying and mapping services		
Engineering design and consulting		
Manufacturing industries		

For the subdivisions Manufacturing, Professional, scientific and technical services and Building cleaning, pest control and other services, the data at the subdivision level includes contributions from other classes outside the BES. Therefore, for these industries the two digit subdivision estimates have to be weighted using the four digit class data for the BES component, and this can be done because the ABS over the last two years has released supplementary tables with data at the subdivision and class level. The data from the survey years for these subdivisions is used to weight the non-survey years by applying the BES proportion in the subdivision's survey year to the other year estimates.

The data is not complete because some industries cannot be separated into the relevant classes from *Australian Industry*. For example, rental of heavy machinery and scaffolding (class 6631) is in subdivision 66 but the data is not available. Also, services such as marketing, legal and financial are important but again not identifiable. Government spending on infrastructure and portfolio investment in departments like health and education is included through the BES supply industries, although any maintenance and work done internally will generally not be included. That also applies in industries like retailing and transport where some (unknown) proportion of work is done in-house. Also, because *Australian Industry* uses tax and business register data, it is the classification of firms to SIC industry classes that fundamentally determines the structure and scope of that data. Needless to say, such classifications are not perfect, particularly in regard to large multi-unit or multi-divisional organizations.

Figure 17. Economic Role of the Australian Built Environment Sector

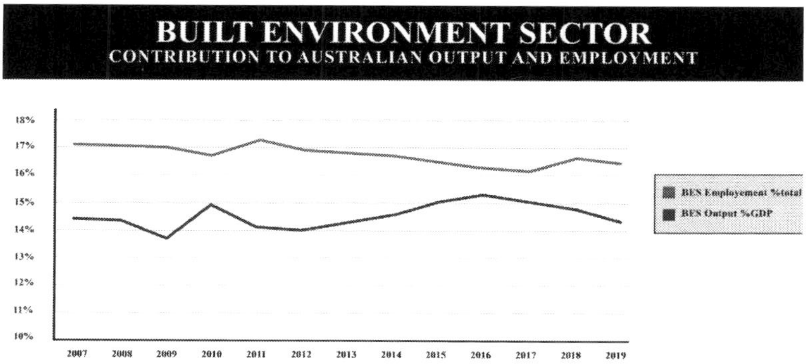

Sources: ABS 8155, ABS 5206, ABS 6202.

Structural changes in the composition of output and employment are the visible signs of creative destruction at work. The industry shares of total BES output and employment have changed over time. The decline of

Manufacturing and decreasing share of Construction has seen the Property and real estate services industry increase its share of IVA from 16 to 26 percent, and share of employment from 15 to 21 percent between 2007 and 2019. The Water and waste and Property and real estate services industries have positive differences between the shares of IVA and employment, reflecting their higher capital requirements and investment in buildings, structures, plant and equipment.

Combining total construction accounts for nearly half of the BES total. That share, however, has fallen from 48 percent of BES IVA to 46 percent between 2007 and 2019, at the same time Construction fell from 53 to 52 percent of total BES employment. Note the share of employment is higher than the share of IVA, due to the labour-intensive nature of Construction Services. After Construction the Property operators and real estate services industry is the largest employer, followed by Professional services. The growth in employment in Professional services is notable, increasing from 9 percent to over 12 percent of BES employment between 2007 and 2019.

Table 14. Industry shares of BES total output and employment 2007 and 2019

		Property & RE	Professional services	Building services	Manufacturing	Water & waste	Construction
Employment	2007	15.4%	9.6%	8.9%	8.8%	2.9%	53.7%
	2019	15.8%	12.2%	9.1%	6.7%	3.1%	52.5%
IVA	2007	20.47%	10.20%	3.46%	10.33%	5.49%	48.69%
	2019	25.61%	9.15%	3.63%	6.66%	7.22%	46.63%

Source: ABS 8155.

The BES IVA and employment data gives the contribution of the individual industries and the structure of the BES, as shown above. This data can also be used to reveal other characteristics of these industries and to make comparisons between them, starting with IVA per person employed. IVA per person employed reflects the capital structure of an industry, which is the investment required for physical capital like machinery and buildings and intellectual capital like patents and processes. The higher the capital requirements, or capital intensity, of an industry the higher the level of IVA per person employed is expected to be because workers with more capital are more productive. Both excavators and shovels require one operator but the former sifts more soil. That effect is seen across the BES, where services like cleaning and construction trades have the lowest level of IVA per person employed, but also have lower capital requirements than the higher IVA per person employed industries of Water and waste, Property and real estate, Professional services, Engineering and Mining.

Figure 18. Output of 16 Australian built environment industries

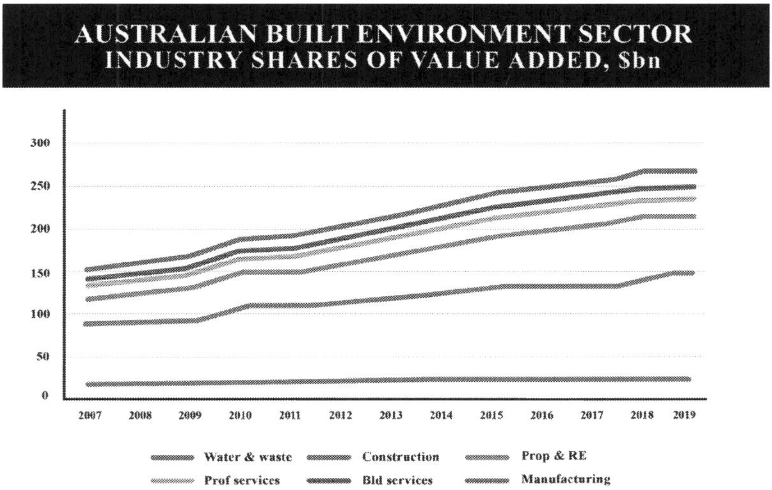

Source: ABS 8155.

Figure 19. Employment in 16 Australian built environment industries

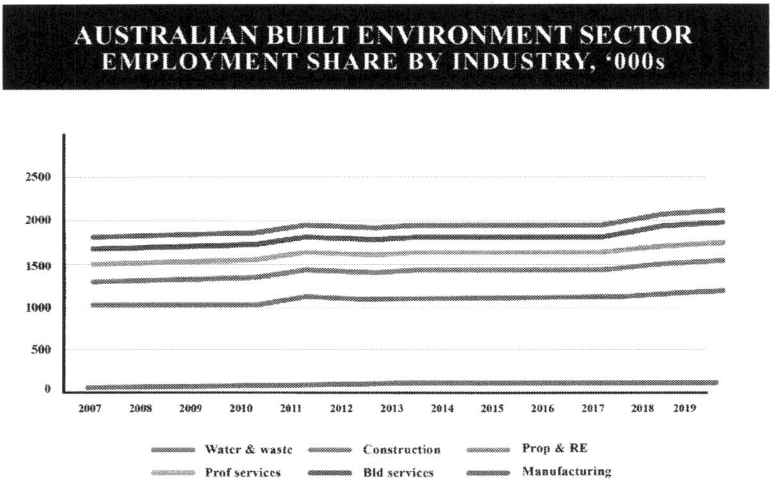

Source: ABS 8155.

As a combination of and employment IVA per person employed looks like a measure of productivity, but while it is indicative that is not the case. Output is not adjusted for price changes and employment is not given in hours worked in *Australian Industry*, therefore the usefulness of IVA per person employed as a proxy for productivity per person is limited. Although these appear to be similar to the output and input data needed to calculate productivity, indexes of output and input are used for productivity analysis. Nevertheless, the small changes seen in Figure 7 reflect the lack of productivity growth many studies have found in construction and related industries.[304]

Figure 20. Output per person employed in 9 industries

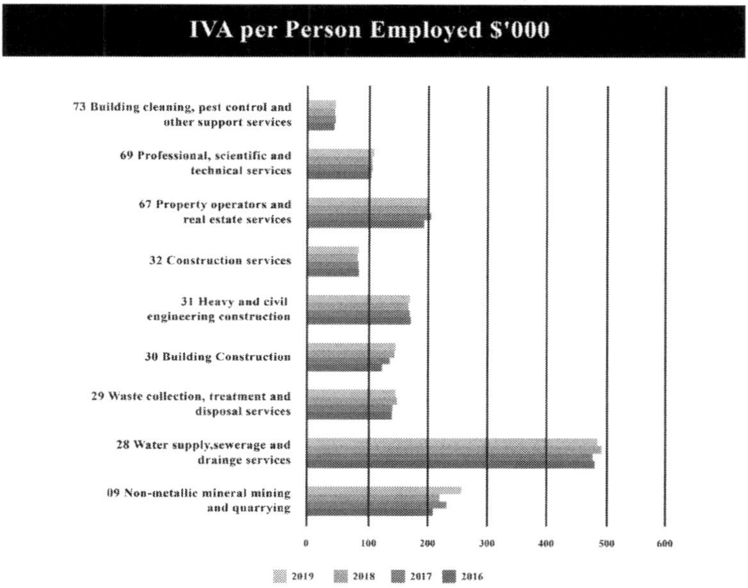

Source: ABS 8155.

The concept of the BES is broad and extensive, so cannot be precise and exact. While the boundaries of industries and markets are important, in practice the SIC classifications are the starting point for the data used.[305] The industries included for Australia have been selected because they clearly have a relationship with construction and maintenance of the built environment. This may not capture every single contribution to the BES, but it does allow the development of a profile of the sector. Measuring the BES provides insight into its relationship to the wider economy, and is relevant to a wide range of policies and issues currently facing the built environment.

Measuring the BES also highlights issues and problems with the construction statistics currently produced by national statistical agencies. Construction economists have been arguing for improvements since Jacque Cannon's 1994 paper 'Lies, dammed lies and construction statistics.' A decade later George Briscoe pointed to the inadequacy of many industry statistics:

> Problems with reliable and accurate data collection and statistical analysis include defining the scope and coverage of the industry; measuring industry outputs and their allocation across different types of activity; identifying construction firms; and measuring capital formation and capital stock, inconsistencies in employment statistics and labour market variables, discrepancies in measuring productivity, and the lack of international comparison.[306]

A Satellite Account and Revising Construction Statistics

The argument is made for two revisions to the construction statistics produced by national statistical agencies. The first revision is for a periodic production of a satellite account for built environment industries. These are extensions of the national accounts published by statistical agencies and are used to group related but separate industries to estimate their combined contribution to the economy. This would produce a built environment sector satellite account similar to the tourism satellite accounts that are found in nine countries.

The second revision concerns the *International Standard Industrial Classification* (ISIC) UN 2008) system of classifying industries. The proposal is to upgrade Residential building, Non-residential building and Engineering construction from their current status as three divisions within the Construction section to new sections in their own right. This requires allocating subcontracted work done by construction trades to the work done (i.e. the economic output) by each section. The discussion here strictly uses the terms given in the ISIC and when referring to an ISIC category they are capitalised, so 'Construction' is an ISIC section, 'Residential building' an ISIC division, and 'construction' is an economic activity. A revision of the existing Construction section would be a major undertaking because previous revisions have only divided a sector into two and no division has been removed.

The problems encountered when trying to measure construction productivity are discussed to highlight shortcomings with the current data available, which has resulted in a distorted and false view of the productivity performance of construction. The discussion then turns to how these problems have been addressed in research done by the US Bureau of Labor Statistics, in particular the two key issues of assigning subcontractor hours to projects and adjusting estimates of output for price changes using deflators. The importance of improving the scope of current construction statistics is discussed, because there have been cases of misunderstanding or misdiagnosis of where or what are problems or issues in construction based on that data that have led to policies that were too often ineffective or irrelevant to large numbers of firms involved in construction of the built environment.

A SATELLITE ACCOUNT FOR THE BUILT ENVIRONMENT SECTOR

How the built environment is created and maintained through project initiation, design, fabrication and construction to operation, repair and maintenance is an ongoing process. The network of firms involved includes construction contractors and subcontractors, property management and real estate services, manufacturers of fittings, finishings, plant and equipment, suppliers of building materials, and professional services. All these firms belong to industries that are part of the process of producing and maintaining the built environment, however,

National agencies collect data and present it in tables following the format given in the *System of National Accounts* (SNA) published by the UN. The national accounts present highly aggregated estimates of expenditure, output and income based on the detailed data collected on the economic activities of households, firms, non-profits and government. That data is collected using the methods, definitions and categories provided in the SNA, ISIC and other publications. Firms and other organizations are assigned ISIC codes on the basis of common characteristics in products, services, production processes and logistics, and collects companies and other organizations into groups with similar characteristics.

Industries as defined by SIC classifications cannot capture all their associated economic activities, and when economic activities involve a range of different industries the contribution of a sector is not obvious, despite its importance. Because the ISIC system puts strict boundaries around an industry, what is included or left out of the definition of an industry determines its extent. However, inclusions and exclusions vary greatly between industries and there are many anomalies. Examples are:

- Health insurance is included in Insurance not in health expenditure
- Retail sales by chemists is included in Pharmaceutical expenditure as well as manufacturing and R&D
- Research is classified to industries not by purpose, and often done by institutions
- Automobile manufacture includes design, Construction does not

The solution to the issues raised by narrow SIC industry definitions is a satellite account that reclassifies expenditures from different industry groupings into a single sector. Satellite accounts have been produced for many sectors that are made up of several industries, such as health, the digital economy, the environment, R&D, the space industry, and infrastructure. They have also been produced for non-profit institutions, volunteering, education and training, and unpaid household activities. They are used to provide more detail on sectors that are not visible in current statistics, following guidelines provided by the SNA for their preparation.[307] The most widely found satellite account is for tourism, so far produced at various times for over 50 countries. This brings together the contributions of industries like travel, accommodation, hospitality, tour operators and entertainment to estimate their total output and employment.[308]

The primary purpose of satellite accounts is to improve policy-making by providing better, more granular data, and demand for satellite accounts has increased as their usefulness has been shown. A 2019 survey by the UN found 80 countries had produced 241 satellite accounts covering over 20 different topics, with 148 of those done since 2000, mainly on health, tourism and the environment. The number produced by country varied from one to 15, the median number of satellite accounts in production was 2 and the average was 4.[309] As a result, there are many guidelines for producing a satellite account available, usually produced through international collaboration, and the methodology has been adapted to a wide variety of sectors.

Figure 21. Number of satellite accounts by sector

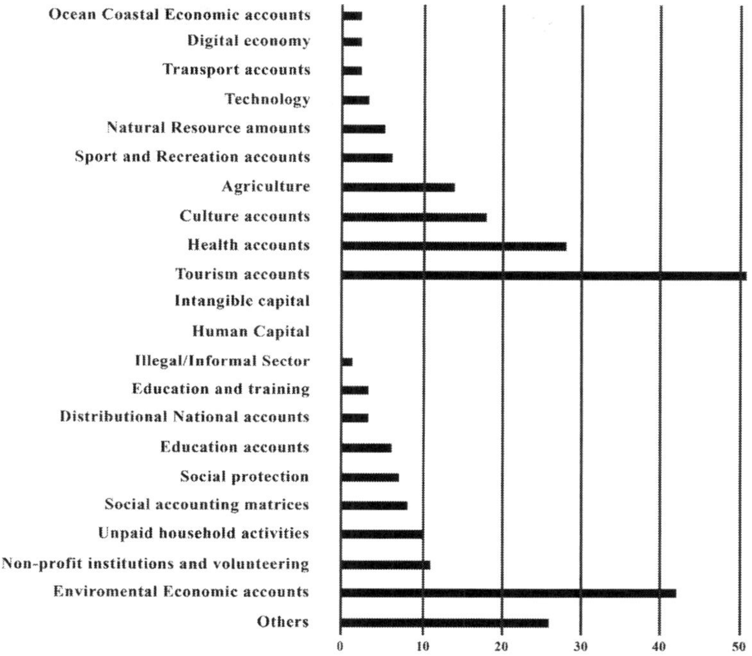

Source: Conference of European Statisticians, 2019: 11. *In-depth review of satellite accounting*, Paris: UNECE.

Preparation of a satellite account requires significant research and development. Different data sources have to be harmonized and measurement challenges met. The OECD published *A System of Health Accounts* in 2000 (updated 2011[310]) after 15 years of development of the concepts and methods needed for a health satellite account, and the US Bureau of Economic Analysis (BEA) worked on their R&D satellite account for over a decade. However, the research is being done and more satellite accounts are being produced, such as the preliminary estimates for The Small Business Economy,[311] and the Space Economy[312] the BEA released in 2020. In 2021 the OECD published the first Working Paper on a Transport satellite account.[313]

A built environment sector satellite account would restrict its scope to relevant activities, and would therefore remain within the production, consumption and asset boundaries of the SNA framework, a type of satellite account known as a thematic account. Some examples of thematic accounts are agriculture, tourism, culture, and sport and recreation. Developing a sector based thematic account involves regrouping, re-arranging and re-packaging existing national accounts data by creating definitions of the economic activities, products, suppliers and users involved.[314] In some cases the national accounts data is supplemented by other sources, such as surveys of household activities or expenditure, that collect data on the use of products and supply of services not otherwise available.

Despite issues of data quality and availability, bringing together the range of industries that contribute to the production, maintenance and management of cities, infrastructure and buildings in a satellite account would improve our understanding of both the sector and the wider economy. For example, urban development and city policies involve significant infrastructure spending, which is often their main focus. However, it is the associated induced industrial, commercial and residential development around the new infrastructure that drives longer-term growth. A satellite account captures that activity.

DIVIDING THE CONSTRUCTION SECTION INTO THREE

The inclusion in the Construction section of the three divisions Residential building, Non-residential building and Engineering construction is a major problem. These should be treated as three separate sections because of the significant differences in their characteristics, but statistical classifications group them together despite these differences. For example, engineering construction uses heavy machinery and equipment and is much more capital intensive than residential building. Most construction productivity research uses aggregate data for construction because, in the statistics published by national statistical agencies, data on employment and hours worked at the level of the three industries is missing or incomplete.

Dividing the Construction section into Residential building, Non-residential building and Engineering construction sections would be a major undertaking, in part because the current classification system and collection methodology, based on surveys and modelling, has been used for a long time and is well known. However, new sources of data have become available and statistical agencies are making more use of digital data from business accounts and tax records, called administrative data. Firms self classify by nominating a SIC code they believe best reflects their primary activities, which is typically not difficult for small and medium size firms. However, a single code is unavoidably inaccurate for large firms with a number of business units that cross boundaries, for example construction contractors that also develop their own buildings, have offsite manufacturing facilities, or do work across the construction divisions.

The most difficult part is assigning work done by Specialized trades to these residential building, non-residential building and engineering construction sections. A method is needed because firms obtain these labour inputs indirectly from subcontractors, so the hours worked by subcontractors is not included in contractor hours worked. This was done for US construction by the Bureau of Labor Statistics (BLS) as part of their ongoing research into construction productivity using data on output, on the number of employees, and on the number of partnerships and proprietorships from the five yearly Census of Construction.[315]

The US Census of Construction shows how much output contractors in each field (such as plumbers or electricians) deliver to each type of construction (such as single or multiple family residential construction} and how much is for new construction, for additions, alterations, and renovations, and for maintenance and repair. The important next step by the BLS researchers was to assign labour hours on the basis of the labour intensity of different construction work. With total labour in new construction known, and total hours worked in Construction is usually collected, labour can be allocated to specific industries. The BLS assumes subcontractor output for additions and alterations are twice as labour intensive (have twice the labour/output ratio) as work done for new construction, and that output for maintenance and repair is three times as labour intensive as output provided to new construction.[316] A 2014 BLS Working paper described the methodology in detail:

> Given these assumptions, we can determine how much of the labor supplied by a particular type of contractor, perhaps carpenters, is allocated to each of these different facets of production. Assume for example that carpenters supply 60 percent of their total output (deliveries to all sectors, not just to home building) to new construction, 20 percent to additions and alterations, and another

> 20 percent to maintenance and repair. In conjunction with the labor/output ratios of 1, 2, and 3, we can state that:
> 0.60 1x + 0.20 2x + 0.20 3x = L where L is the total labor input employed by the carpenters. [317]

In this example L = 0.375, so if 80 percent of carpenter labour to new construction is single family home building 0.80 times 0.375 means 30 percent of total carpenter labour is supplied to single family home building. This can be done for plumbers, roofers and other contractors supplying single family housing, then adding the contributions from them determines the total amount of labour provided by contractors.

Traditionally statistical agencies relied on surveys to collect data from construction firms, but the variety of special trades and the sheer number of micro and small firms makes getting adequate samples from surveys difficult and expensive. Although assumptions can always be questioned and tested, they are needed if a model based on labour intensity is to be used for allocating hours worked by subcontractors to the three main divisions. Statistical agencies use models in their estimates of output of construction work done, based on percentages of different types of buildings in total work done. Developing a model for allocating hours worked to the type of construction can be done if the required data is available.

The BLS is in the fortunate position of being able to use the Census of Construction, many countries do not do regular surveys or have such detailed data. There are also other data sources available to the BLS. The five year gap between Census years means it is necessary to interpolate data, and here the BLS has access to: annual data on employment from the BLS Current Employment Statistics; data on employee hours in each industry from the Current Employment Statistics; on the number of partnerships and proprietorships using data from Census on Non-Employer Statistics; and on the average weekly hours of partners and proprietors in construction from the BLS Current Population Survey.

Few countries have the range and depth of data sets as the US, in another example the NAICS classification for Manufacturing has a six digit level where four digits is the norm in other countries. Nevertheless, a program can be done to revise construction classifications so Residential building, Non-residential building and Engineering construction become sections and subcontracted work and labour hours are included in those new sections. Statistical agencies should be given the opportunity. Economic and industry policies that affect construction would benefit from more disaggregated data, which would also allow better targeting and monitoring of those policies. Turning the current divisions into sections would also provide a new perspective on the characteristics and performance of the three different industries that are currently grouped in Construction, and in particular address the widespread but false impression that construction has a productivity problem.

Construction Deflators and Productivity

The tools, techniques, components and materials used in modern construction can be seen on every building site. As anyone who works in construction knows, they have greatly increased the productivity of workers, but that increase in productivity cannot be seen in construction statistics. For decades there has been little or no growth in construction productivity as measured by national statistical agencies.

Many reasons have been given for this stagnant growth of productivity are various and include volatility of demand, fragmentation and the number of small firms, the one-off nature of projects, the high labour intensity of residential building, poor economies of scale, limited competition, regulatory impediments, a lack of innovation, poor management, low levels of capital investment and of skills.[318] The rates of technology adoption and diffusion has always been an issue.[319] Alternatively the BLS and other researchers suggest it is possible the statistics on the levels of productivity in the industry might be faulty,[320] or the measures misleading,[321] or the comparisons inappropriate.[322] These issues will be taken in turn.

INAPPROPRIATE COMPARISONS

Construction productivity is a soft target. The industry was labelled as backward by Woudhuysen and Abley in their 2004 book *Why is Construction so Backward?* The UK industry was compared, badly, with manufacturing in the 1998 Egan Report and described as inefficient, fragmented and adversarial in the 1994 Latham Report.[323] These UK advocates of industrialized building pressed for construction to adopt similar production practices to manufacturing, particularly car manufacturing.[324]

However, while there are factory made structures and components, the number of standard buildings is limited. The opportunities for standardised construction products and assemblies are much more widespread and onsite production is organised around those standard parts and materials. Manufacturing, in contrast, is organised around standardised products and continuous production runs.

Table 15. UK construction and manufacturing compared by size of firm

	Construction					**Manufacturing**				
Employee size band	No. of firms	%	Total turnover £m	%	Average turnover per firm £k	No. of firms	%	Total turnover £m	% %	Average turnover per firm £k
1 – 9	328,936	94	107,076	35	325	107,177	78	29,003	5	271
10 – 49	17,197	5.3	57,991	19	3,372	21,523	16	56,798	10	2,639
50 – 249	2,132	0.6	52,284	17	24,523	6,466	5	118,166	21	18,275
Over 249	333	0.1	91,771	29	275,588	1,419	1	366,645	64	258,383
All	348,598	100	309,123	100	887	136,585	100	570,611	100	4,178

Source: Meikle, J. and de Valence, G. 2022. Construction products and producers: One industry or three, in Best, R. and Meikle, J. (eds.) *Describing Construction: Industries, projects and firms*, London: Taylor and Francis. Data from ONS Annual Business Survey 2018.

Lessons from other industries and their production methods and processes can be useful and informative, however, comparing performance between industries is very difficult without adjustments to make the subjects comparable. With the comparison between construction and manufacturing, the problem is that both measures are averages of extremely varied activities collected by the standard industrial classification system. This makes useful comparisons between the two difficult.

Although the SIC groups all construction firms into a single category, that is for statistical convenience based on conventions developed originally for classifying manufacturing. The exclusion of design from construction output while included in manufacturing and the inclusion of R&M in construction but not in manufacturing is the result.[325] Another is the view of construction as a single industry, producing and maintaining buildings and structures, despite their many different types and the differences in the producers and processes used in their delivery.

Table 16. Comparing UK construction and vehicle manufacture[326]

	Construction	**Vehicle manufacture**
Total turnover	GBP 309 billion	GBP 61 billion
Number of enterprises	348,598	10,660
Total employment	1,562,000	86,000
Turnover per worker	GBP 197,902	GBP 707,965
Capital intensity	low	high
Labour intensity	high	low
location	mobile site based	fixed factory based
Product design	mostly in professional services	mostly by vehicle manufacturers
Standard products	very small proportion	majority of production
Work on existing products	substantial proportion	R&M excluded

Source: Meikle, J. and de Valence, G. 2022. Construction products and producers: One industry or three, in Best, R. and Meikle, J. (eds.) *Describing Construction: Industries, projects and firms*, London: Taylor and Francis. Data from ONS Annual Business Survey 2018.

These differences in clients, projects, firms and output all support the idea that construction is a collection of industries, not one single industry, albeit with overlaps among them. Measures like the number of employees and value added per employee vary across them.[327] The broad categories used in the SIC of residential building, non-residential building and civil engineering construction are different enough to call for different types of materials, contractors and implementation processes. However, the industry classification 'Construction' does not separate the data on employment and output so the productivity statistics are misleading.

MISLEADING MEASURES

Researchers in the US Bureau of Labour Statistics (BLS) produced productivity estimates for four types of construction that show the importance of separating the different industries/divisions in Construction for measuring productivity.[328] Their data and analysis was a significant advance on the aggregate construction productivity estimates that have become familiar. The four industries are:

- Single-family residential construction
- Multiple-family housing construction
- Industrial building construction
- Highway, street, and bridge construction

As the figures below show, productivity fell in single-family residential and multiple-family housing construction, but rose in industrial and highway, street, and bridge construction. Between 2007 and 2020 overall productivity was flat because these rises and falls balanced out. Also, 2007 was the peak of a US business cycle, followed by a recession from December 2007 to June 2009 that 'had both immediate and lasting impacts on the construction industries.' Two of the four industries show clear and strong productivity growth, which remains positive with subcontractor labour included and grew fastest in industrial building construction. The trends in output, hours worked and labour productivity in the four industries were distinct and different.

The data and methods commonly used to estimate the level of productivity in the industry are misleading because they are incomplete, and give a false picture of construction productivity. The BLS research required combining four different datasets to show declining productivity in residential building but increasing productivity in non-residential building and construction.

Figure 22. US Construction labour productivity

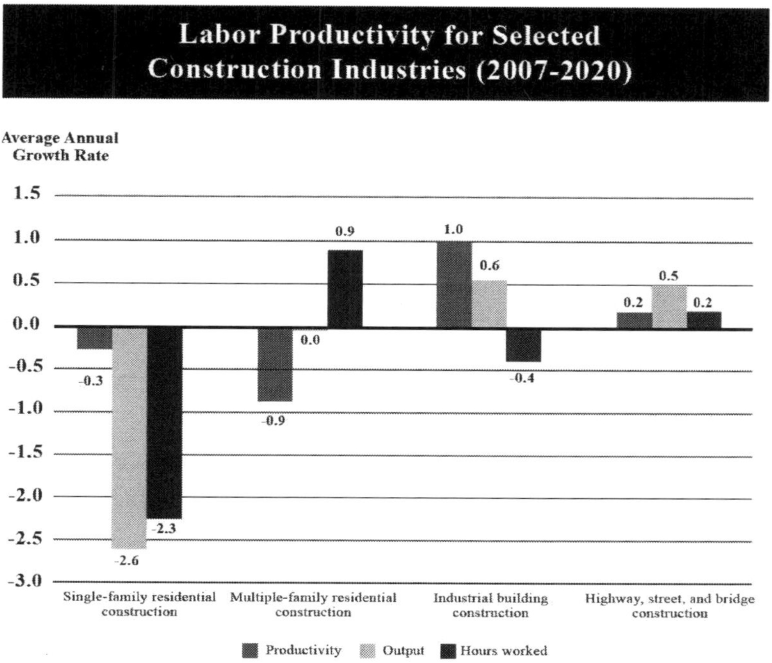

Figure 23. US Construction labour productivity for four industries

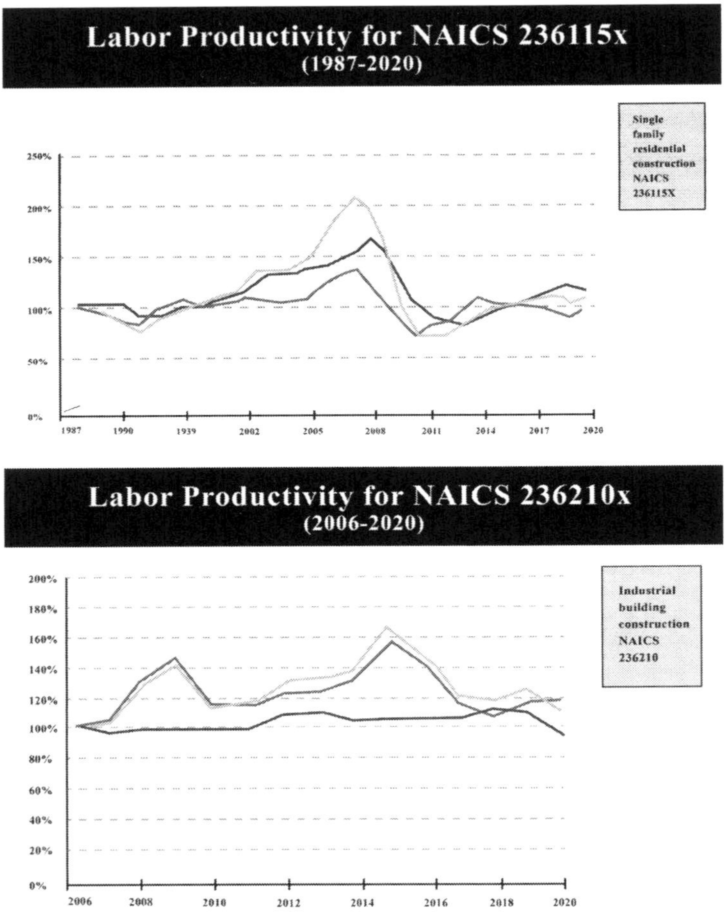

Labor Productivity for NAICS 236116x
(1987-2020)

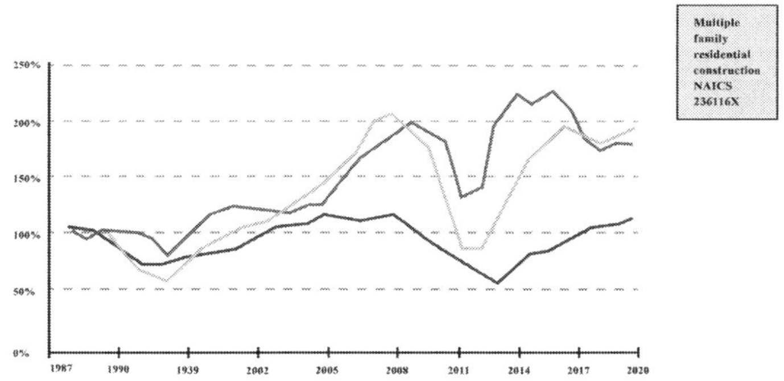

Labor Productivity for NAICS 237310
(2002-2020)

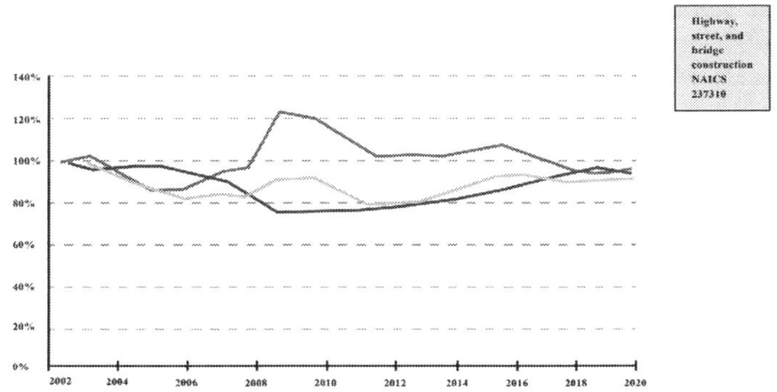

FAULTY STATISTICS

A more technical problem with construction statistics is the method used to adjust industry output for changes in prices of materials and labour to find changes in the quantity of output of completed buildings and structures. This deflation of output is typically done with input price indexes or producer price indexes. The problem with input price indexes is they assume a constant relationship between input and output over time,[329] so there is an assumption of no change in productivity which means that, if productivity is increasing, input price indexes will be upwardly biased.

The main reason for the low rate of measured productivity growth in construction are the deficiencies found in construction deflators. Productivity estimates require both a measure of labour inputs, such as hours worked or people employed, and a measure of output, usually industry value added (the difference between total revenue and total costs) adjusted for changes in prices of materials and labour. That deflated measure of output is known as real construction value added.

If real construction value added is underestimated due to the deflators used, construction productivity has also been understated. Thus the graphs of flatlining construction productivity, despite the obvious improvements in materials, tools and techniques over the last few decades. There is a downward bias to output estimates when there is no adjustment for quality changes in buildings and structures.

As the energy efficiency and quality of finishes has improved, and as the share of building costs due to mechanical and electrical services has increased over time (providing greater amenity), the deflators used have not been adjusted to take these trends into account. In effect, the deflators assume there has been no change in the quality of buildings, and their inability to capture quality changes in the buildings and structures delivered by the construction industry has adversely affected the measurement of productivity.

Another problem is the application of a deflators to the diverse range of buildings and structures, and differences in quality and function between them. The application of a single deflator to heterogeneous goods, especially durable goods, overlooks differences in age and function. This problem becomes more severe with long-life assets like buildings and structures.

The BLS productivity estimates for four construction sub-industries used four deflators from different government databases. Their research addresses the problem with new data: 'The main difficulty is that buildings differ widely in their characteristics and features. Similarly, the nature of the underlying terrain varies widely among construction projects. Consequently, economists, both in general and within the BLS productivity program, have found it exceptionally difficult to develop reliable output price deflators to convert observed revenues into meaningful measures of output growth over time. Good output price deflators are therefore the key to more accurate measures of productivity growth in construction.'

The researchers then say: 'we examine only those industries in which the deflators exactly match the industry boundaries. Previous work generally looked at the total construction sector. Since the many new deflators now available did not exist then, these prior studies had to use the single-family housing deflator and an associated cost index to deflate production in most or all of construction.'[330]

The BLS research addressed the main reason for the low rate of measured productivity growth in construction, which are the deficiencies found in construction deflators. There is a downward bias to output estimates because there is no adjustment for quality changes in buildings and structures. If real construction value added is underestimated due to the deflators used, construction productivity has also been understated. Thus, flatlining construction productivity despite the obvious improvements in materials, tools and techniques over the last few decades.

There is an extensive literature on deflators, the problems of deflation, and the effects on estimates of construction output of commonly used deflators. The issues raised by the use of price indexes for deflation have not been solved to date, and appear to have no simple, or readily available solutions. The fundamental problem is that the deflator used to adjust for price changes will systematically overstate the rate at which prices increase and underestimate growth in output if indices for labour and material costs are used instead of output price indices (which are generally not available).

The US productivity estimates for four construction sub-industries used four different deflators, providing high quality estimates of real construction value added per hour worked in those industries, including subcontractor hours. The BLS research improves on previous research by using appropriate output deflators to develop measures of productivity growth, therefore their measures are more reliable because the deflators are specifically designed for each industry. Their data and analysis was a significant advance on the widely used aggregate construction estimates that led to a false idea of industry productivity. It demonstrates both how more accurate statistics can be produced, using the extensive data available to the BLS, and why separating the three divisions of Residential building, Non-residential building and Engineering construction and upgrading them to ISIC sections of their own is necessary.

Bibliography

ABS 2009. *Information Paper: Introduction of revised international statistical standards in the Australian Tourism Satellite Account*, Australian Bureau of Statistics, Cat. no. 5249.0.55.002.

ABS 2004. *Capitalising Research and Development in the National Accounts*. National Accounts Research Section. Australian Bureau of Statistics.

ABS, 1998. *Construction Industry Survey 1996-97*. Australian Bureau of Statistics, Cat. No. 8772.0.

Acemoglu, D. 2002. Directed Technical Change. *Review of Economic Studies*. 69: 781-809.

Acemoglu, D., P. Aghion, L. Bursztyn, and D. Hemous. 2012. The Environment and Directed Technical Change. *American Economic Review*. 102(1): 131-166.

Acemoglu D. and Restrepo, P. 2019. Artificial Intelligence, Automation, and Work, in Agrawal, A., Gans, J. and Goldfarb, A. (eds.) *The Economics of Artificial Intelligence: An Agenda*, London: The University of Chicago Press. 197-236.

AEGIS, 1999. *Mapping the Building and Construction Product System in Australia*. Australian Expert Group on Industry Studies. Canberra: Department of Industry, Science and Resources.

Agrawal, A., Gans, J. and Goldfarb, A. 2018. *Prediction Machines: The Simple Economics of Artificial Intelligence*, Harvard, Mass.: Harvard Business Review Press.

Aghion, P., A. Dechezleprêtre, D. Hemous, R. Martin, and J. Van Reenen. 2016. Carbon Taxes, Path Dependency and Directed Technical Change: Evidence from the Auto Industry. *Journal of Political Economy*. 124: 1-51.

Aiginger, K. and Rodrik, D. 2020. Rebirth of Industrial Policy and an Agenda for the Twenty-first Century, *Journal of Industry, Competition and Trade*, 20:189–207.

Allen, R. 2009. *The British Industrial Revolution in Global Perspective*,

Cambridge: Cambridge University Press.
An, P., Martinez, S., Al-Hussein, M. and Ahmad, R. 2020. BIM-based decision support system for automated manufacturability check of wood frame assemblies, *Automation in Construction,* 111 130-65.
Ando, F, Y., Funo, S., Furusaka, S., Matsudome, S., Matsumura, S., Murakami, S., Yashiro, T., Yoshida, T., Groak, S., Drewer, S., Gann, D., Ive, G. and Meikle, J. 1992. A comparative study on technological change in the house building industries in Japan and the United Kingdom (1). *Housing Research Foundation Annual Report, 18,* 379–392.
Andrews, D., C. Criscuolo and P. N. Gal, 2015. *Frontier Firms, Technology Diffusion and Public Policy: Micro Evidence from OECD Countries,* OECD Productivity Working Papers, 2015-02, Paris: OECD.
ARUP, 2020. *Designing with Digital Fabrication*, London.
Arthur, W. B. 2009. *The Nature of Technology: What it is and How it Evolves*, London: Penguin.
Arthur, W. B. 1988. Competing Technologies, Increasing Returns and 'Lock-In' by Small Historical Events, *Economic Journal*, 99, 394, 116-131.
Bainbridge, L. 1983. Ironies of automation. *Automatica, 19* (6), 775-779.
Bamigboye, G. O., Davies, I., Nwanko, C., Michaels, T., Adeyemi, G. and Ozuor, O. 2019. Innovation in construction materials - A review, *IOP Conf. Ser.: Mater. Sci. Eng.* 640 012070
BCA, 2016. *BIM Essential Guide for DfMA* (Design for Manufacturing and Assembly), Building and Construction Authority, Singapore.
BCG, 2017. *Digital in Engineering and Construction: The Transformative Power of Building Information Modeling*, Boston Consulting Group.
Beamish, T. D and Biggart, N. W. 2012. The role of social heuristics in project-centred production networks: insights from the commercial construction industry, *Engineering Project Organization Journal*, 2:1-2, 57-70.
Best, R., Langston, C. and de Valence, G. (eds.) 2003. Building in Value: Workplace Strategies and Facilities Management, Butterworth-Heinemann, Oxford.
Best, R. and de Valence, G. (eds.) 2002. Building in Value: Design and Construction, Butterworth-Heinemann, Oxford.

Best, R. and de Valence, G. (eds.) 1999. Building in Value: Pre-Design Issues, Arnold Publishers, London.

Best, R. and Meikle, J. (eds.) 2022. *Describing Construction: Industries, projects and firms*, London: Taylor and Francis.

Best, R. and Meikle, J. (eds.) 2019. *Accounting for Construction: Frameworks, productivity, cost and performance.* Abingdon: Routledge.

Best, R. and Meikle, J. (eds.) 2015. *Measuring Construction: Prices, output and productivity*. London: Taylor & Francis.

Bianco, J. L., Fuchs, S., Parsons, M. and Riberinho, M. J. 2018. *Artificial Intelligence: Construction Technology's Next Frontier*, McKinsey & Co.

Bijker, W. E., Hughes, T. P. and Pinch, T. 1987. *The Social Construction of Technological Systems: New Directions in the Sociology and History of Technology*, Cambridge, Mass.: MIT Press.

Bloom, N., Van Reenen, J. and Williams, H. 2019. A toolkit of policies to promote innovation. *Journal of Economic Perspectives*, *33*(3), 163-84.

Bloom, N., Sadun, R. and Van Reenen, J. 2016. *Management as a Technology,* working paper 22327, Cambridge, Mass: National Bureau of Economic Research

Bloom, N. and Van Reenen, J. 2010. Why Do Management Practices Differ across Firms and Countries? *Journal of Economic Perspectives*, 24(1), 203–224.

Bon, R. (ed.) 2000. *Economic Structure and Maturity: Collected Papers in Input-Output Modelling and Applications,* Aldershot: Ashgate.

Bonvillian, W. B. and Weiss, C. 2015. *Technological Innovation in Legacy Sectors*, Oxford University Press USA.

Bowley, M. 1966. *The British building industry- Four Studies in Response and Resistance to Change,* Cambridge: Cambridge University Press.

Bowley, M. 1960. *Innovation in Building Materials*, London: Gerald Duckworth.

Brand, S. 1994. *How Buildings Learn*, New York: Viking Presss.

Braun, I. and Joerges, B. 1994. How to recombine large technical systems: the case of European organ transplantation, in Summerton, J. (ed.) *Changing Large Technical Systems*, Boulder (CO): Westview.

BRE, 2001. *Current practice and potential uses of prefabrication,* Building Research Establishment, Watford, U.K

BRESNAHAN, T. F. and Trajtenberg, M. 1995. General purpose technologies 'Engines of growth'? *Journal of Econometrics.* 65 (1): 83.

BRESNEN, M. 1990. *Organizing Construction: Project Organization and Matrix Management.* London: Routledge.

BRESNEN, M. and Marshall, N. 2001. Understanding the diffusion and application of new management ideas in construction, *Engineering Construction and Architectural Management*, 8 (6), 335-45.

BRISCOE, G. 2006. How useful and reliable are construction statistics? *Building Research & Information, 34*(3), 220-229.

BRÖCHNER, J. 2022. Construction economics in antiquity, in Ofori, G. (ed.) *Research Companion to Construction Economics*, Cheltenham: Edward Elgar. 86-103.

BRÖCHNER, J. 2011. Developing construction economics as industry economics, in de Valence, G. (ed.). *Modern Construction Economics,* London: Taylor and Francis. 14-28.

BROCKMANN, C. 2022. *Construction Microeconomics*, New York: Wiley-Blackwell.

BRYNJOLFSSON, E. and McAfee, A. 2014. *The Second Machine Age: Work, Progress and Prosperity in a Time of Brilliant Technologies*, New York: W. W. Norton & Co.

CABINET OFFICE 2016. *Government Construction Strategy 2016-2020,* Cabinet Office, London.

CANNON, J., 1994. "Lies and Construction Statistics", *Construction Management and Economics*, Vol. 12, No. 4, 307-12.

CARASSUS, J., Andersson, N., Kaklauskas, A., Lopes, J., Manseau, A., Ruddock, L. and de Valence, G. 2006. Moving from production to services: A built environment cluster framework, *International Journal of Strategic Property Management*, 10, 169-84.

CALLENDER, G. 2003. A Short History of Procurement, in P. Nagel, *Supply Chain Management: A Procurement Perspective,* Melbourne: Hargreen Publishing, 2-9.

CANNON, J. 1994. Lies and construction statistics. *Construction Management and Economics, 12*(4), 307-313.

CARROLL, P. 2010. Does regulatory impact assessment lead to better policy? *Policy & Society*, 29:2, 113-122.

CARASSUS, J., Andersson, N., Kaklauskas, A., Lopes, J., Manseau, A., Ruddock, L. and de Valence, G. 2006. Moving from production to services: A built environment cluster framework, *International Journal of Strategic Property Management*, **10**, 169-84.

Cassimatis, P.J. 1969. *Economics of the Construction Industry*, National Industrial Conference Board, New York.

Chan, T. C. 2019. Cost ratios and technology choice, in Best, R. and Meikle, J. (eds.) *Accounting for Construction: Frameworks, productivity, cost and performance,* London: Taylor & Francis. 129-154.

Chancellor, W., Abbott, M., and Carson, C. 2019. Measuring construction industry activity and productivity: The impact of the shadow economy. In R. Best & J. Meikle (Eds.), *Accounting for Construction: Frameworks, Productivity, Cost and Performance.* New York: Routledge. 74-86

Chang, H-J. and Andreoni, A. 2020. Industrial Policy in the 21st Century, *Development and Change,* 51(2): 324–351.

Chaoran Chen, C. 2020. Technology adoption, capital deepening, and international productivity differences, *Journal of Development Economics*, Volume 143, 102388.

Charlton, J., Kelly, K., Greenwood, D. and Moreton, L. 2021. The complexities of managing historic buildings with BIM, *Engineering, Construction and Architectural Management*, Vol. 28 No. 2, pp. 570-583.

Clark, L. 1992. *Building Capitalism – Historical Change and the Labour Process in the Production of the Built Environment*, London: Routledge.

Cody, J. 2003. *Exporting American Architecture 1870-2000*, London: Routledge.

Coe, C.K. 1989. *Public Financial Management.* Englewood Cliffs, NJ: Prentice Hall.

Connaughton, J. and Meikle, J. 2013. The changing nature of UK construction professional service firms. *Building Research and Information*, **41** (1), 95-109.

Cooney, E.W. 1955. The Origins of the Victorian Master Builders, *Economic History Review*, 2nd Series, No. 8, 167-76.

Corrado, C. and Hulten, J, 2010: 11. Measuring intangible capital: How do you measure a "technological revolution"? *American Economic Review,* 100: 99–104.

Cousineau, L. and Miura, N. 1998. *Construction Robots: The Search for*

New Building Technology in Japan. ASCE Press.

COUTARD, O. 1999. The evolving forms of governance of large technical systems, in Coutard, O. (ed.) *The Governance of Large Technical Systems*, London: Routledge.

CRC 2007. *Offsite manufacture in Australia: Final report.* Research Project No: 2005-004-C-03, Cooperative Research Centre for Construction Innovation, Brisbane, Australia.

CRUZ, S. C. and Teixeira, A. A. 2010. The evolution of the cluster literature: Shedding light on the regional studies–regional science debate. *Regional Studies*, 44(9), 1263–1288.

DACY, D.C. 1965. Productivity and price trends in construction since 1947. *Review of Economics and Statistics*, 47(4), 406-411.

DAVID, P. A. 1990. The Dynamo and the Computer: An Historical Perspective on the Modern Productivity Paradox, *American Economic Review.* 80, 355 - 361.

DAVID, P. A. 1985. Clio and the Economics of QWERTY, *American Economic Review*, 75(2) 332-337.

DAVIS, H. 2006. *The Culture of Building*, Oxford: Oxford University Press.

DE VALENCE, G. 2022. On the measurement and characteristics of construction firms in theory and practice, in Best, R. and Meikle, J. (eds.) *Describing Construction: Industries, projects and* firms, London: Taylor and Francis.

DE VALENCE, G. 2022. Output and employment in Australian built environment industries 2007-19, in Best, R. and Meikle, J. (eds.) *Describing Construction: Industries, projects and firms*, London: Taylor and Francis.

DE VALENCE, G. 2022. The nature of construction economics, in Ofori, G. (ed.) *Research Companion to Construction Economics*, Cheltenham: Edward Elgar. 61-85.

DE VALENCE, G. 2019. Reframing construction within the Built Environment Sector, *Engineering, Construction and Architectural Management*, 26 (5) 740-745.

DE VALENCE, G. (ed.). 2011. *Modern Construction Economics: Theory and application*, London: Taylor & Francis.

DE VALENCE, G. 2010. Innovation, Procurement and Construction Industry Development, *Australasian Journal of Construction Economics and Building*, 10 (4) 50-59.

de Valence, G. 2001. Defining an industry: what is the size and scope of the Australian building and construction industry? *The Australian Journal of Construction Economics and Building*, Vol. 1, No. 1, 53-65.

de Valence, G. 1997. Construction Industry Reform Strategies in Australia: CIDA and the CPSC, First International Conference on Construction Industry Development, National University of Singapore, Vol. 1, pp. 124-31.

de Valence, G. and Abbott, M. 2015. A review of the theory and measurement techniques of productivity in the construction industry, in Best, R. and Meikle, J. (eds.) *Measuring Construction: Prices, Output and Productivity*, London: Taylor & Francis, pp. 205-23.

de Valence, G. and Algeo, C. 2016. Transformational Procurement Model for Managing Megaprojects. ANZAM 2016 Conference, Sydney.

Destafano, T., Kneller, R., and Timmins, J. 2018. Broadband Infrastructure, ICT use and firm performance: Evidence for UK Firms, *Journal of Economic Behavior & Organization,* 155(C), pp 110-139.

Dosi, G. 1988. Sources, Procedures, and Microeconomic Effects of Innovation, *Journal of Economic Literature*, 26(3), 1120-1171.

Dosi, G. 1982. Technological paradigms and technological trajectories. A suggested interpretation of the determinants and directions of technical change, *Research Policy*, 11(3):147-162.

Dosi, G., Freeman, C. and Fabiani, S. 1994. The Process of Economic Development: Introducing Some Stylized Facts and Theories on Technologies, Firms and Institutions, *Industrial and Corporate Change,* 3(1), 1-45.

Dubois, A. and Gadde, L-E. 2002. The construction industry as a loosely coupled system: implications for productivity and innovation, *Construction Management & Economics*, 20 (7), 621-31.

Eccles, R.1981. Bureaucratic versus craft administration: the relationship of market structure to the construction firm. *Administrative Science Quarterly, 26*(3), 449-469.

Edgerton, D. 2007. *The Shock of the Old: Technology and Global History since 1900*. Oxford: Oxford University Press.

Egan, J. 1998. *Rethinking construction,* London: Department of the Environment, Transport and the Regions.

Farmer, M. 2016. *Modernise or die*, London: Construction Leadership

Council.

Fernie, S., Leiringer, R. and Thorpe, T. 2006. Rethinking change in construction: a critical perspective. *Building Research & Information*, **34**(2), 91-103.

Fitch, J.M. 1973. *American Building: The Historical Forces that Shaped It*, New York: Shocken Books.

Forman, C., Goldfarb, A. and Greenstein, S. 2018. How Geography Shapes – and is Shaped by – the Internet, in (ed.) Clark, G. Feldman, M. Gertler, M. and Wojcik, D. *The New Oxford Handbook of Economic Geography,* Oxford University Press. 269-285.

Fraumeni, B. M. and Okubo, S. 2005. R&D in the National Accounts: A First Look at Its Effect on GDP. In *Measuring Capital in the New Economy*, Corrado, C., Haltiwanger, J. and Sichel, D. (eds.) University of Chicago Press. 275–321.

Gann, D. M. 2003. Innovation in the built environment, *Construction Management & Economics*, 21, 553–5.

Gann, D. M. 2001. Putting academic ideas into practice: technological progress and the absorptive capacity of construction organizations, *Construction Management & Economics*, 19 (3), 321-30.

Gann, D. 1996. Construction as a Manufacturing Process – similarities and differences between Industrialized housing and car production in Japan, *Construction Management & Economics* (14) pp 437 – 450.

Gann, D. M. and Salter, A. 2000. Innovation in project-based, service-enhanced firms: the construction of complex products and systems, *Research Policy*, 29, 955-72.

Gershenfeld, N. 2012. How to Make Almost Anything: The Digital Fabrication Revolution. *Foreign Affairs,* 91(6), pp. 43–57.

Gershenfeld, N. 2017. The Science, and The Roadmap, in Gershenfeld, N., Gershenfeld, A. and Cutcher-Gershenfeld, J. *Designing Reality: How to Survive and Thrive in the Third Digital Revolution*, in New York: Basic Books. 95-116, and 159-182.

Gibb, A.G. 1999. *Off-site fabrication - pre-assembly, prefabrication and modularisation*, Whittles Publishing Services.

Gibb, A.G.F and Isack, F. 2003. Re-engineering through pre-assembly: client expectations and drivers. *Building Research & Information*, 31(2), 146-160.

Goldin, C, and Katz, L. 1998. The Origins of Technology-Skill Complementarity. *Quarterly Journal of Economics* 113 (3): 693–732.

Green, B. 2019. The challenges of measuring British construction output. In R. Best and J. Meikle (Eds.), *Accounting for Construction: Frameworks, Productivity, Cost and Performance,* pp. 46-73. New York: Routledge.

Green, S.D. 2011. *Making sense of Construction Improvement*, Wiley-Blackwell, Oxford.

Green, S.D., Fernie, S., Weller, S. 2005. Making sense of supply chain management: a comparative study of aerospace and construction, *Construction Management & Economics*, 23(6), 579-593.

Green, S.D. and May, S.C. 2003. Re-engineering construction: going against the grain, *Building Research & Information,* 31(2), 97-106

Greenstein, S. 2015. *How the Internet Became Commercial: Innovation, Privatization, and the Birth of a new Network,* Princeton University Press.

Greenwood, D. 2019. The adoption of building information modelling as a measure of international construction innovation' in Gruneberg, S. (ed.) 2019. *Global Construction Data*, London: Taylor & Francis. 111 -125.

Greif, A., Milgrom, P. and Weingast, B. 1994. Coordination, Commitment, and Enforcement: The Case of the Merchant Guild. *Journal of Political Economy,* 102(4): 912–50.

Groak, S. 1992. *The Idea of Building*, London: Spon.

Gruneberg, S. 2020. *A Strategic Approach to the UK Construction Industry,* London: Taylor and Francis.

Gruneberg, S. (ed.) 2019. *Global Construction Data*, London: Taylor & Francis.

Gruneberg, S. and Francis, N. 2019. *The Economics of Construction*, Newcastle upon Tyne: Agenda Publishing.

Gruneberg, S and Ive, G. 2000. *The Economics of the Modern Construction Firm*, London: Macmillan.

Gruszka, A., Jupp, J. R. and de Valence, G. 2017. *Digital Foundations: How Technology is Transforming Australia's Construction Sector,* Sydney: Startup Australia.

Hall, D. M., Whyte, J. K. and Lessing, J. 2019. Mirror-breaking strategies to enable digital manufacturing in Silicon Valley construction firms: a comparative case study, *Construction Management & Economics*, doi:10.1080/01446193.2019.1656814

Harvey, J.H. 1975. *Medieval Craftsmen*, London: B.T. Batsford.

Hausmann, R., and Hidalgo, C. A. 2011. The network structure of economic output, *Journal of Economic Growth* 16:309–342.

Helpman. E. and Trajtenberg, J. 1990. Diffusion of General Purpose Technologies, in Helpman, E. (ed.), *General Purpose Technologies and Economic Growth*, Cambridge Mass.: MIT Press. 85-119.

Hobday, M. 2000. The project-based organization: an ideal for managing complex products and systems? *Research Policy*, 29, 871-93.

Hughes, T. P. 2004. *Human-Built World: How to Think about Technology and Culture*, Chicago: University of Chicago Press.

Hughes, T. P. 1989. *American Genesis: A Century of Invention and Technological Enthusiasm 1870-1970*, Chicago: University of Chicago Press.

Hughes, T. P. 1987. The evolution of large technological systems, in *The Social Construction of Technological Systems: New Directions in the Sociology and History of Technology*, W. E. Bjiker, T. P. Hughes, and T. J. Pinch (eds.), Cambridge, Mass.: MIT Press.

Hughes, T. P. 1983. *Networks of Power: Electrification in Western Society, 1880–1930.* Baltimore: Johns Hopkins University Press.

Huxtable, A. L. 2008. *On Architecture: Collected Reflections on a Century of Change*, New York: Walker Publishing Company.

Ickes, B.W. and Samuelson, L. 1987. Job transfers and incentives in complex organizations: Thwarting the ratchet effect. *RAND Journal of Economics* 18, no. 2 (Summer): 275–86.

Ive, G. and Gruneberg, S. L. 2000. *The Economics of the Modern Construction Sector*, London: Macmillan.

Ive, G., Gruneberg, S., Meikle, J., & Crosthwaite, D. 2004. *Measuring the competitiveness of the UK construction industry* (Vol. 1). London: Department of Trade and Industry.

Jackson, D. 2018. *Data Cities,* Lund Humphries, London.

Jackson, K. 1987. *Crabgrass Frontier: The Suburbanization of the United States.* Oxford: Oxford University Press.

Jahren, P. 2011. *Concrete: History and Accounts,* Trondheim: Tapir Academic Press.

Jiang, R., Wu, C., Lei, X., Shemery, A., Hampson, K.D. and Wu, P. 2021.

Government efforts and roadmaps for building information modelling implementation: lessons from Singapore, the UK and the US, *Engineering, Construction and Architectural Management*, https://doi.org/10.1108/ECAM-08-2019-0438

Jin, W. and McElheran, K. 2018. Economies before Scale: Learning, Survival, and Performance of Young Plants in the Age of Cloud Computing. SSRN Electronic Journal: https://papers.ssrn.com/sol3/papers.cfm?abstract_id=3112901

Jorgenson, D. and Landefeld, S. 2006. Blueprint for Expanded and Integrated U.S. Accounts: Review, Assessment, and Next Steps, in *A New Architecture for the U.S. National Accounts*, Jorgenson, D., Landefeld, S. and Nordhaus, W. (eds.). NBER Studies in Income and Wealth, Vol. 60. University of Chicago Press.

Kagel, J. H. and Levin, D. 1986. The Winner's Curse and Public Information in Common Value Auctions. *American Economic Review*, 894–920.

Kamien, M. L. and Schwartz, N. L. 1982. *Market Structure and Innovation*, Cambridge: Cambridge University Press.

Kaseman, K. and Graser, K. 2020. Digital fabrication in the construction sector, in Sawhney, A., Riley, M. and Irizarry, J. (eds.) *Construction 4.0: An Innovation Platform for the Built Environment*, Abingdon: Routledge.

Katz, M. L., and Shapiro, C. 1985. Network Externalities, Competition, and Compatibility, *American Economic Review*, 75(3): 424-440.

Katz, M. L., and Shapiro, C. 1986. Technology Adoption in the Presence of Network Externalities, *Journal of Political Economy*, 94(4): 822-841.

Katz, M. L., and Shapiro, C. 1992. Product Introduction with Network Externalities, *Journal of Industrial Economics*, 40(1): 55-83.

Kingsford, P.W. 1973. *Builders and Building Workers*, London: Edward Arnold.

Kelly, K. 2016. *The Inevitable: Understanding the 12 Technological Forces That Will Shape Our Future,* New York: Random House.

Knoop, G. and Jones, G.P. 1933. *The Medieval Mason*, Manchester: Manchester University Press.

Khoshnevis, B. 2004. Automated construction by contour crafting related robotics and information technologies, *Journal of Automation in*

Construction. 13 (1) 5-19.

Koc, E., Pantazis, E., Soibelman, L. and Gerber, D. 2020. Emerging trends and research directions, in Sawhney, A., Riley, M. and Irizarry, J. (eds.) *Construction 4.0: An Innovation Platform for the Built Environment*, Abingdon: Routledge. 460 -75.

Landes, D. S. 1972. *The Unbound Prometheus: Technological Change from 1750 to the Present*, Cambridge: Cambridge University Press.

Lavikka, R., Kallio, J., Casey, T. and Airaksinen, M. 2018 Digital disruption of the AEC industry: technology-oriented scenarios for possible future development paths, *Construction Management and Economics*, 36:11, 635-650.

Latham, M. 1994. *Constructing the team.* Final Report of the Government/ Industry Review of Procurement and Contractual Arrangements in the UK Construction Industry, London: HMSO.

Lee, G. and Bormann, A. 2020. BIM policy and management, *Construction Management and Economics*, 38:5, 413-419.

Leibwicz, B. D. 2017. Effects of urban land-use regulations on greenhouse gas emissions, *Cities*, 70, 135-152.

Lena-Acebo, F. J., and García-Ruiz, M. E. 2019. The FABLAB Movement: Democratization of Digital Manufacturing. In A. Guerra Guerra (Ed.), *Organizational Transformation and Managing Innovation in the Fourth Industrial Revolution.* IGI Global. pp. 125-142.

Lim, M. K., Xiong, W. and Wang, C. 2021. Cloud manufacturing architecture: a critical analysis of its development, characteristics and future agenda to support its adoption, *Industrial Management & Data Systems.*

Lipsey, R.G., Carlaw, K. I. and Bekar, C. T. 2005. *Economic Transformations: General Purpose Technologies and Long-term Economic Growth*, Oxford: Oxford University Press.

Loulakis, M.C. 2003. *Design-build for the public sector.* New York: Aspen.

Manuelli, R.E. and Seshadri, A. 2014. Frictionless Technology Diffusion: The Case of Tractors, *American Economic Review. 104 (4): 1368-91.*

McElheran, K. 2015. Do Market Leaders Lead in Business Process Innovation? The Case(s) of E-Business Adoption, *Management Science*, 61(6), 1197-1216.

McKellar, E. 1999. *The Birth of Modern London: The Development and Design of the City 1660- 1720,* Manchester: Manchester University Press.

McKinsey, 2017. *Reinventing construction through a productivity revolution*, McKinsey Global Institute.

Meikle, J. 2009. *Thinking Big: A History of Davis Langdon,* London: Black Dog Publishing.

Meikle, J. and de Valence, G. 2022. Construction products and producers: One industry or three, in Best, R. and Meikle, J. (eds.) *Describing Construction: Industries, projects and firms*, London: Taylor and Francis.

Meikle, J. and Gruneberg, S. 2015. Measuring and comparing construction internationally. In Best, R. and Meikle, J. (eds.) *Measuring Construction: Prices, Output and Productivity*, Abingdon: Routledge.

Miao, Q. and D. Popp. 2014. Necessity as the Mother of Invention: Innovative Responses to Natural Disasters. *Journal of Environmental Economics and Management*. 68(2): 280- 295.

Misa, T. J. 2009. History of Technology, in *Companion to the Philosophy of Technology*, J. K. B. Olsen, S. A. Pedersen, and V. F. Hendricks (eds.), Chichester, UK: Wiley-Blackwell, 7-17.

Mitchell, M. 2019. *Artificial Intelligence: A Guide for Thinking Humans,* New York: Farrar, Straus, and Giroux.

Morris, P.W.G. 2013. *Reconstructing Project Management,* Oxford: Wiley-Blackwell.

Mowery, D.C. and Rosenberg, N. 1998. *Paths of Innovation: Technological Change in 20th Century America*, Cambridge: Cambridge University Press.

Munn, L. 2022. *Automation is a Myth,* Stanford University Press.

Murray, A. and Kulakov, A. 2019. Make/buy decisions in international construction industry firms, in Gruneberg, S. (ed.) *Global Construction Data*, London: Taylor & Francis. 149-166.

Murray, M. and Langford, D. 2003. *Construction Reports 1944-98,* Oxford: Wiley-Blackwell.

Myers, D. 2017. *Construction Economics: A new approach,* 4th Ed. London: Spon Press.

National Audit Office, 2001. *Modernising Construction,* London: National Audit Office.

NOAILLY, J. 2012. Improving the Energy Efficiency of Buildings: The Impact of Environmental Policy on Technological Innovation. *Energy Economics*. 34: 795-806.

OECD, 2002. *Frascati Manual 2002: Proposed Standard Practice for Surveys on Research and Experimental Development,* Paris: OECD Publications.

OFORI, G. 1990. *The Construction Industry: Aspects of its Economics and Management,* Singapore: Singapore University Press.

OGILVIE, S. 2014. The Economics of Guilds, *Journal of Economic Perspectives*, 28:4, pp. 169-92.

OKUBO, S., Robbins, C., Moylan, C., Sliker, B., Schultz, L. and Mataloni, S. 2006. R&D Satellite Account: Preliminary Estimates, Bureau of Economic Analysis.

ONS, 2007. *Standard Industrial Classification of Economic Activities.* Office of National Statistics, London.

PASQUIRE, C.L. and Gibb, A.G.F. 2002. Considerations for assessing the benefits of standardisation and pre-assembly in construction, *Journal of Financial Management of Property and Construction*, 7(3), 151–61.

PATH, 2004. *Technology Roadmap: Advanced panelised construction, 2003 Progress Report.* Partnership for Advanced Technology in Housing (PATH), Department of Housing and Urban Development, Office of Policy Development and Research, Washington, D.C.

PARK, W. and Chapin, W. 1992. *Construction Bidding: Strategic pricing for profit*, 2nd ed. New York: Wiley

PEARCE, D. 2003. *The Social and Economic Value of Construction: The Construction Industry's Contribution to Sustainable Development*, London: nCrisp.

PETERS, T. F. 1996. *Building the Nineteenth Century*, Cambridge, Mass.: MIT Press.

PETROSKI, H. 1997. *Remaking the World: Adventures in engineering*, Random House, New York.

PFAMMATTER, U. 2008. *Building the Future: Building Technology and Cultural History from the Industrial Revolution until Today.* Munich: Prestel Verlag.

POLAT, G., Arditi, D., Ballard, G., Mungen, U. 2006. Economies of on-site versus off-site fabrication of Rebar, *Construction Management &*

Economics (24) 1185-1198.

Popp, D. 2019. Environmental Policy and Innovation: A Decade of Research. CESifo Working Paper No. 7544, SSRN: https://ssrn.com/abstract=3352908

Popp, D., R. Newell and A.B. Jaffe 2010. Energy, the Environment, and Technological Change. In *Handbook of the Economics of Innovation: vol. 2*, Bronwyn Hall and Nathan Rosenberg, eds., Academic Press/Elsevier, 873-937.

Porter, M. 1985. *Competitive Advantage*, New York: Free Press.

Porter, M. 1980. *Competitive Strategy*, New York: Free Press.

Powell, C. G. 1980. *An Economic History of the British Building Industry 1815-1979*, London: The Architectural Press.

Powell, W. 1990. Neither market nor hierarchy: network forms of organization. *Research in Organizational Behavior*, 12: 295–336.

Queensland Government 2018. Digital Enablement for Queensland Infrastructure - Principles for BIM Implementation, State of Queensland, Queensland.

Quezada, G., Bratanova, A., Boughen, N. and Hajowicz, S. 2016. *Farsight for Construction: Exploratory scenarios for Queensland's construction industry to 2036*, CSIRO, Australia.

Rabeneck, A. 2011. The invention of the building industry in Britain, *ArtefaCTos*, 4:1, 93-121.

Rich, N. and Malik, N 2019. *International Standards for Design and Manufacturing: Quality Management and International Best Practice*, London: Kogan Page.

Rodrik, D. and Stantcheva, S. 2021. Fixing capitalism's good jobs problem, *Oxford Review of Economic Policy*, 37: 824-837.

Rosenburg, N., 1982. *Inside the Black Box: Technology and Economics*, Cambridge: Cambridge University Press.

Rosenfielde, S. and Mills, D. 1979. Is construction technologically stagnant? In J.E. Lange & D.Q. Mills, (eds.) *The Construction Industry*, D.C. Heath and Company, Lexington MA.

Ruddock, L. and Ruddock, S. 2009. The scope of the construction sector: determining its value, in Ruddock, L. (ed.) *Economics for the Modern Built Environment*, London: Routledge.

Runeson, G. 2000. *Building Economics*, Deakin: Deakin University Press.

Runeson, G. and de Valence, G. 2013. International Construction: From

Transnational to Global, AUBEA Conference, Auckland.

Runeson, G. and de Valence, G. 2008. The New Construction Industry, in Ruddock, L. (ed.) *Economics for the Modern Built Environment*, London: Taylor and Francis, 199-211.

Sanjayan, J. G., Nazari, A. and Nematollahi, B. 2019. *3D Concrete Printing Technology - Construction and Building Applications.* London: Elsevier.

Satoh, A. 1995. *Building in Britain: The Origins of a Modern Industry*, London: Scholar Press.

Sawhney, A., Riley, M. and Irizarry, J. (eds.) 2020. *Construction 4.0: An Innovation Platform for the Built Environment*, Abingdon: Routledge.

Schumpeter, J. 1942. *Capitalism, Socialism and Democracy*, New York: Harper.

Sichel, D. 2022. The Price of Nails Since 1695: A Window into Economic Change, *Journal of Economic Perspectives, Volume 36, Number 1: 125–150.*

Simmons, R. 2015. Constraints on evidence-based policy: insights from government practices, *Building Research & Information*, 43:4, 407-419.

Sovacool, B. K., Lovell, K. and Ting, M. B. 2018. Reconfiguration, Contestation, and Decline: Conceptualizing Mature Large Technical Systems, *Science, Technology, & Human Values*, Vol. 43(6) 1066-1097.

Squicciarini, M. and Asikainen, A-L. 2011. A value chain statistical definition of construction and the performance of the sector, *Construction Management and Economics*, 29 (7), 671-693.

Straub, H. 1952. *A History of Civil Engineering*, Cambridge, MA: MIT Press.

Strassman, P. 1970. The construction sector in economic development. *Scottish Journal of Political Economy*, 17(3): 391–409.

Stinchcombe, A.L. 1959. *Bureaucratic* and *craft administration* of production: A Comparative Study, *Administrative Science Quarterly*, 4, 168-87.

Strategic Forum, 2002. *Accelerating Change, Rethinking Construction.* London:

Summerson, J. 1945. *Georgian London,* edited by Colvin, H. 2003, Yale

University Press.
Summerton, J. 1994. The systems approach to technological change, in Summerton, J. (ed.) *Changing Large Technical Systems*, Boulder: Westview.
Sutton, J. 1999. *Technology and Market Structure*, Cambridge Mass.: MIT Press.
Sveikauskas, L., Rowe, S., Mildenberger, J., Price, J. and Young, A. 2018. Measuring productivity growth in construction, *Monthly Labor Review*, U.S. Bureau of Labor Statistics.
Sveikauskas, L., Rowe, S., Mildenberger, J., Price, J. and Young, A. 2016. Productivity growth in construction, *Journal of Construction Engineering and Management*, vol. 142, no. 10.
Sveikauskas, L., Rowe, S., Mildenberger, J., Price, J. and Young, A. 2014. Productivity growth in construction, Bureau of Labor Statistics. Working Paper 478.
Syverson, C. 2019. Macroeconomics and Market Power: Context, Implications, and Open Questions, *Journal of Economic Perspectives*, 33, 3, 23–43.
Syverson, C. 2008. Markets: Ready-Mixed Concrete, *Journal of Economic Perspectives*, 22, 1, 217–233.
Taylor, M. 2010. A definition and valuation of the UK offsite construction sector, *Construction Management & Economics*, 28:8, 885-896.
Thornton, R. 2002. *The Houses That Sears Built: Everything You Ever Wanted to Know About Sears Catalogue Homes,* Alton, Illinois Gentle Beam Publications.
Trajtenberg, M. 2019. Artificial Intelligence as the Next GPT: A Political-Economy Perspective, in Agrawal, A., Gans, J. and Goldfarb, A. (eds.) *The Economics of Artificial Intelligence: An Agenda*, London: The University of Chicago Press. 175-188.
Triplett, J.E. 1983. Concepts of Quality in Input and Output Price Measures: A Resolution of the User-Value Resource-Cost Debate. In Foss, M.F. (ed) *The U.S. National Income and Product Accounts: Selected Topics*, NBER Studies in Income and Wealth, Vol. 47. Chicago: National Bureau of Economic Research.
Turin, D. 2003. Building as a process, *Building Research & Information*, 31:2, 180-187.
Turin, D. (ed.) 1975. *Aspects of the Economics of Construction.* Goodwin,

London
Turin, D. 1969. *The Construction Industry: Its Economic Significance and Its Role in Development,* New York: UNIDO. UN Statistical Division, 1968. A System of National *Accounts*, Dept. of Economic and Social Affairs, Statistical Office, New York.
United Nations Environment Programme. 2021: 59. *Global Status Report for Buildings and Construction: Towards a Zero-emission,* Efficient and Resilient Buildings and Construction Sector. Nairobi.
UN, 2008. *International Standard Industrial Classification of all Economic Activities (ISIC).* New York: United Nations Publications.
UN, 2009. *System of National Accounts 2008.* New York: United Nations
Vanoli, A. 2005, *A History of National Accounting*, Amsterdam: IOS Press.
Varian, H. 2019. Artificial Intelligence, Economics, and Industrial Organization, in Agrawal, A., Gans, J. and Goldfarb, A. (eds.) *The Economics of Artificial Intelligence: An Agenda*, London: The University of Chicago Press. 175-188.
Varian, H. 2014. Beyond Big Data, *Business Economics*, 49(1): 27-31.
Vollebergh, H.R.J. and E. van der Werf. 2014. The Role of Standards in Eco-Innovation: Lessons for Policymakers. *Review of Environmental Economics and Policy*. 8: 230–248.
Warwick, K. 2013. Beyond Industrial Policy: Emerging Issues and New Trends, *OECD Science, Technology and Industry Policy Papers*, No. 2, OECD Publishing, Paris.
Wei, W., Zhou, F. and Liang, P.F. 2020. Product platform architecture for cloud manufacturing. *Advances in Manufacturing* 8, 331–343.
Weitzman, M. 1980. The "ratchet principle" and performance incentives. *Bell Journal of Economics* 11: 302–8.
Weitzman, M. 1976. The new Soviet incentive model. *Bell Journal of Economics,* 7: 251-57.
WEF, 2017. *Future Scenarios and Implications for the Construction Industry,* World Economic Forum and the Boston Consulting Group, Geneva.
WEF, 2016. *Shaping the Future of Construction: A Breakthrough in Mindset and Technology*, World Economic Forum and the Boston Consulting Group, Geneva.
Wilkinson, T. 2014. *Bricks and Mortals: Ten great buildings and the people they made*, New York: Bloomsbury Press.

WINCH, G. 2006. Towards a theory of construction as production by projects, *Building Research & Information,* 34:2, 154-163.

WOLSTENHOLME, A. 2009. *Never Waste a Good Crisis: A review of progress since rethinking construction and thoughts for our future*, London: Construction Excellence.

WOMACK, J. P., Jones, D. T. and Roos, D., 1990. *The machine that changed the world: Based on the Massachusetts Institute of Technology 5-million dollar 5-year study on the Future of the Automobile*, Rawson Associates, Toronto, Collier Macmillan New York.

WOND, T. and Macaulay, M. 2010. Evaluating local implementation: An evidence-based approach. *Policy & Society*, 29:2, 161-169.

WOUDHUYSEN, J. Abley, I. 2004. *Why is Construction so Backwards*, Chichester: Wiley-Academy.

YU, M. and Ive, G. 2015. Review of construction cost and price indices in Britain, in Best, R. and Meikle, J. (eds.) *Measuring Construction: Prices, output and productivity*, Oxford: Taylor and Francis, pp. 79-112.

ZOLAS, N. et al. 2020. *Advanced Technologies Adoption And Use By U.S. Firms: Evidence From The Annual Business Survey,* Nikolas Zolas, Zachary Kroff, Erik Brynjolfsson, Kristina McElheran, David N. Beede, Cathy Buffington, Nathan Goldschlag, Lucia Foster and Emin Dinlersoz. National Bureau of Economic Research, Cambridge, MA Working Paper

Index

M

N

O

P

Q

R

S

T

U

Endnotes

Introduction

[1] Hughes, T. P. 2004: 12. *Human-Built World: How to Think about Technology and Culture*, Chicago: University of Chicago Press.

[2] Connaughton, J. and Meikle, J. 2013. The changing nature of UK construction professional service firms. *Building Research and Information*, 41 (1), 95-109.

[3] Murray, A. and Kulakov, A. 2019. Make/buy decisions in international construction industry firms, in Gruneberg, S. (ed.) *Global Construction Data*, London: Taylor & Francis. 149-166.

[4] Powell, C.G. 1980. *An Economic History of the British Building Industry 1815-1979*, London: The Architectural Press.

[5] Misa, T. J. 2009. History of Technology, in *Companion to the Philosophy of Technology*, J. K. B. Olsen, S. A. Pedersen, and V. F. Hendricks (eds.), Chichester, UK: Wiley-Blackwell, 7-17.

Many studies are focused on the invention and diffusion of specific technologies, among other examples discussed here are tractors in agriculture, electricity generation, concrete and wooden frame housing. Another approach is a case study of a demonstration or flagship project, such as a new seed variety, power plant or building.

[6] Samuelson's 1947 *Foundations of Economic Analysis* was the template for the many economics textbooks that followed. In a 2009 interview at the age of 94, his advice was 'Have a very healthy respect for the study of economic history, because that's the raw material out of which any of your conjectures or testings will come'. https://www.theatlantic.com/politics/archive/2009/06/an-interview-with-paul-samuelson-part-two/19627/

[7] Lipsey, R.G., Carlaw, K. I. and Bekar, C. T. 2005. *Economic Transformations: General Purpose Technologies and Long-term Economic Growth*, Oxford: Oxford University Press.

[8] Schumpeter, J. 1942. *Capitalism, Socialism and Democracy*, New York: Harper.

[9] Sawhney, A., Riley, M. and Irizarry, J. (eds.) 2020. *Construction 4.0: An Innovation Platform for the Built Environment*, Abingdon: Routledge.

[10] The Appendix discusses this idea and provides details on the economic role of built environment industries.

[11] The implications of Thomas Hughes' industry life cycle for construction are analysed in chapter 3. Hughes, T. P. 1987. The evolution of large technological systems, in *The Social Construction of Technological Systems: New Directions in the Sociology and History of Technology*, W. E. Bjiker, T. P. Hughes, and T. J. Pinch (eds.), Cambridge, Mass.: MIT Press.

[12] Although wildly optimistic about the future, the key points Kelly makes about data, cloud-based AI and unbundling of products and services are important. Kelly, K. 2016. *The Inevitable: Understanding the 12 Technological Forces That Will Shape Our Future,* New York: Random House.

[13] Gann, D. M. and Salter, A. 2000. Innovation in project-based, service-enhanced firms: the construction of complex products and systems, *Research Policy*, 29, 955-72. Where the initial version of this figure is attributed to a 1992 report comparing UK and Japanese housebuilding (Ando et al. 1992).

[14] Edgerton, D. 2007. *The Shock of the Old: Technology and Global History since 1900.* Oxford: Oxford University Press.

[15] Charlton, J., Kelly, K., Greenwood, D. and Moreton, L. 2021. The complexities of managing historic buildings with BIM, *Engineering, Construction and Architectural Management*, Vol. 28 No. 2, pp. 570-583.

[16] Bresnen, M. 1990. *Organizing Construction: Project Organization and Matrix Management.* London: Routledge.

Hobday, M. 2000. The project-based organization: an ideal for managing complex products and systems? *Research Policy*, 29, 871-93.

[17] Dosi, G. 1982. Technological paradigms and technological trajectories. A suggested interpretation of the determinants and directions of technical change, *Research Policy*, 11(3):147-162.

[18] Trajtenberg, M. 2019. Artificial Intelligence as the Next GPT: A Political-Economy Perspective, in Agrawal, A., Gans, J. and Goldfarb, A. (eds.) *The Economics of Artificial Intelligence: An Agenda*, London: The University of Chicago Press. 175-188.

[19] Mitchell, M. 2019. *Artificial Intelligence: A Guide for Thinking Humans,* New York: Farrar, Straus, and Giroux.

[20] Pfammatter, U. 2008. *Building the Future: Building Technology and*

Cultural History from the Industrial Revolution until Today. Munich: Prestel Verlag.

[21] Many examples of worker resistance in the nineteenth and twentieth centuries can be found in Munn, L. 2022. *Automation is a Myth,* Stanford University Press.

[22] Rodrik, D. and Stantcheva, S. 2021. Fixing capitalism's good jobs problem, *Oxford Review of Economic Policy,* 37: 824-837.

[23] Schumpeter, J. 1942. *Capitalism, Socialism and Democracy*, New York: Harper.

[24] Lipsey, R.G., Carlaw, K. I. and Bekar, C. T. 2005: 13. *Economic Transformations: General Purpose Technologies and Long-term Economic Growth*, Oxford: Oxford University Press. They include two organizational GPTs: mass production and the factory system; and lean production and the Toyota system.

[25] Sutton, J. 1999. *Technology and Market Structure*, Cambridge Mass.: MIT Press.

[26] Peters, T. F. 1996. *Building the Nineteenth Century*, Cambridge, Mass.: MIT Press. Chapter 3 has more detail on these dimensions and examples of them at work.

[27] Hughes, T. P. 1989. *American Genesis: A Century of Invention and Technological Enthusiasm 1870-1970*, Chicago: University of Chicago Press.

[28] Agrawal, A., Gans, J. and Goldfarb, A. 2018. *Prediction Machines: The Simple Economics of Artificial Intelligence*, Harvard, Mass.: Harvard Business Review Press.

[29] Gershenfeld, N. 2012. How to Make Almost Anything: The Digital Fabrication Revolution. *Foreign Affairs,* 91(6), pp. 43–57.

[30] Aiginger, K. and Rodrik, D. 2020. Rebirth of Industrial Policy and an Agenda for the Twenty-first Century, *Journal of Industry, Competition and Trade*, 20:189–207.

[31] Hausmann, R., and Hidalgo, C. A. 2011. The network structure of economic output, *Journal of Economic Grow*th 16:309–342.

Industrial Revolutions and Creative Destruction

[32] Arthur, W. B. 2009: 194. *The Nature of Technology: What it is and How it Evolves*, London: Penguin.

[33] de Valence, G. 2019. Reframing construction within the Built

Environment Sector, *Engineering, Construction and Architectural Management*, 26 (5) 740-745.

[34] Lipsey, R.G., Carlaw, K. I. and Bekar, C. T. 2005. *Economic Transformations: General Purpose Technologies and Long-term Economic Growth*, Oxford: Oxford University Press. They list two dozen GPTs since 9000BCE.

[35] David, P. A. 1990. The Dynamo and the Computer: An Historical Perspective on the Modern Productivity Paradox, *American Economic Review*. 80, 355 - 361.

[36] Acemoglu D. and Restrepo, P. 2019. Artificial Intelligence, Automation, and Work, in Agrawal, A. Gans, J. and Goldfarb, A. (eds.) *The Economics of Artificial Intelligence: An Agenda*, London: The University of Chicago Press. 197-236.

[37] '... the innovation that has shaped recent economic growth is not an autonomous event that falls like manna from heaven. Nor is it a result of R&D and ICT investments alone. Instead, a surge of new ideas (technological or otherwise) is linked to output growth through a complex process of investments in technological expertise, product design, market development, and organizational capability.' Corrado, C. and Hulten, J, 2010: 11. Measuring intangible capital: How do you measure a "technological revolution"? *American Economic Review,* 100: 99–104.

[38] Bloom, N., Sadun, R. and Van Reenen, J. 2016. *Management as a Technology,* working paper 22327, Cambridge, Mass: National Bureau of Economic Research

[39] Bresnahan, T. F. and Trajtenberg, M. 1995. General purpose technologies 'Engines of growth'? *Journal of Econometrics.* 65 (1): 83.

[40] Helpman. E. and Trajtenberg, J. 1990: 86. Diffusion of General Purpose Technologies, in Helpman, E. (ed.), *General Purpose Technologies and Economic Growth*, Cambridge Mass.: MIT Press. 85-119.

[41] Hughes, T. P. 1987: 47. The evolution of large technological systems, in *The Social Construction of Technological Systems: New Directions in the Sociology and History of Technology*, W. A technological system is hard to define is because it has fuzzy boundaries and takes many forms: electricity grids and railways have networks; navigation, telecommunications and air traffic use interconnected nodes; postal systems use existing networks; others such as water, sewerage and drainage or city partnerships are loosely linked networks; some are geographically large,

some are local; some are narrow, some broad. A system can function in a decentralized way, through combinations of local arrangements.

[42] Arthur, W. B. 2009. *The Nature of Technology: What it is and How it Evolves*, London: Penguin.

[43] The scope of technology given by Edgerton (2007) is machines, structures and processes available to users. That technology is used to create physical changes is well-known, Landes in *The Unbound Prometheus* (1972) mentions it frequently, and one of engineering historian Henry Petroski's books was called *Remaking the World* (1997).

[44] Hausmann, R. and Hidalgo, C. A. 2011. The network structure of economic output, *Journal of Economic Growth* 16:309–342.

[45] Kamien, M. L. and Schwartz, N. L. 1982. *Market Structure and Innovation*, Cambridge: Cambridge University Press.

[46] Pfammatter, U. 2008. *Building the Future: Building Technology and Cultural History from the Industrial Revolution until Today*. Munich: Prestel Verlag.

[47] In *Capitalism, Socialism and Democracy* (1942) the chapter on creative destruction is only five pages, however three chapters discuss the process of innovation. Over several decades from *The Theory of Economic Development* in 1911 to his monumental *Business Cycles* in 1938 Schumpeter developed a consistent body of thought.

[48] Rosenburg, N, 1982: 159. *Inside the Black Box: Technology and Economics*, Cambridge: Cambridge University Press.

[49] Helpman. E. and Trajtenberg, J. 1990. Diffusion of General Purpose Technologies, in Helpman, E. (ed.), *General Purpose Technologies and Economic Growth*, Cambridge Mass.: MIT Press. 85-119.

[50] Dosi, G. 1988. Sources, Procedures, and Microeconomic Effects of Innovation, *Journal of Economic Literature*, 26(3), 1120-1171.

[51] Sutton, J. 1999. *Technology and Market Structure*, Cambridge Mass.: MIT Press.

[52] For an analysis of this process in three industries see Mowery, D.C. and Rosenberg, N. 1998. *Paths of Innovation: Technological Change in 20th Century America*, Cambridge: Cambridge University Press.

For the impact of the Toyota system on the global automobile industry see Womack, J. P., Jones, D. T. and Roos, D., 1990. *The Machine that Changed the World,* Rawson Associates, Toronto, Collier Macmillan New York.

[53] Gann, D. M. 2003. Innovation in the built environment, *Construction Management & Economics*, 21, 553–5.

[54] Bonvillian, W. B., and Weiss, C. 2015: 45. *Technological Innovation in Legacy Sectors*, New York: Oxford University Press. 'Paradigm incompatible technologies are by definition disruptive, challenging the business and organizational models that dominate the particular legacy sector. Thus, unless they can locate a market niche in which to begin, they must assault the legacy citadel.' p. 60.

[55] Bloom, N., Van Reenen, J. and Williams, H. 2019. A toolkit of policies to promote innovation. *Journal of Economic Perspectives*, *33*(3), 163-84.

[56] Dosi, G., Freeman, C. and Fabiani, S. 1994. The Process of Economic Development: Introducing Some Stylized Facts and Theories on Technologies, Firms and Institutions, *Industrial and Corporate Change*, 3(1), 1-45.

[57] An example is Gruneberg, S. 2020. *A Strategic Approach to the UK Construction Industry*, London: Taylor and Francis.

[58] Arthur, W. B. 2009. *The Nature of Technology: What it is and How it Evolves*, London: Penguin. He argues the three key characteristics of advanced technologies such as AI are combination, recursiveness, and phenomena. In combination technologies allow AI enabled platforms to generate value. The value for users increases as predictions improve with increasing processing and networking power which creates recursiveness, because AI consists of technologies that are made up of other technologies. Arthur's phenomena is data driven learning

Continuity and Change in Construction After 1800

[59] Davis, H. 2006: 259. *The Culture of Building*, Oxford: Oxford University Press.

[60] For examples and details see Bröchner, J. 2022. Construction economics in antiquity, in Ofori, G. (ed.) *Research Companion to Construction Economics*, Cheltenham: Edward Elgar. 86-103.

[61] Coe, C.K. 1989: 87. *Public Financial Management*. Englewood Cliffs, NJ: Prentice Hall.

[62] Straub. *A History of Civil Engineering*, 1952; and Morris, P.W.G. 2013: 14. *Reconstructing Project Management*, Oxford: Wiley-Blackwell. Both based on Plutarch.

[63] Cited by Callender, G. 2003: 3. A Short History of Procurement, in P. Nagel, *Supply Chain Management: A Procurement Perspective,* Melbourne: Hargreen Publishing,

[64] Knoop, G. and Jones, G.P. 1933. *The Medieval Mason*, Manchester: Manchester University Press.

[65] Kingsford, P.W. 1973. *Builders and Building Workers*, London: Edward Arnold.

[66] Greif, A., Milgrom, P. and Weingast, B. 1994. Coordination, Commitment, and Enforcement: The Case of the Merchant Guild. *Journal of Political Economy,* 102(4): 912–50.

[67] Harvey, J.H. 1975. *Medieval Craftsmen*, London: B.T. Batsford.

[68] McKellar, E. 1999: 86. *The Birth of Modern London: The Development and Design of the City 1660- 1720,* Manchester: Manchester University Press.

[69] McKellar, E. 1999: 83. *The Birth of Modern London: The Development and Design of the City 1660- 1720,* Manchester: Manchester University Press.

[70] Summerson, J. 1945:9. *Georgian London,* edited by Colvin, H. 2003, Yale University Press.

[71] Ogilvie, S. 2014: 169. The Economics of Guilds, *Journal of Economic Perspectives*, 28:4, pp. 169-92.

[72] Harvey, J.H. 1975: 53.. *Medieval Craftsmen*, London: B.T. Batsford.

[73] Ogilvie, S. 2014: 180. The Economics of Guilds, *Journal of Economic Perspectives*, 28:4, pp. 169-92.

Greif, A., Milgrom, P. and Weingast, B. 1994. Coordination, Commitment, and Enforcement: The Case of the Merchant Guild. *Journal of Political Economy,* 102(4): 912–50.

[74] McKellar, E. 1999: 71. *The Birth of Modern London: The Development and Design of the City 1660- 1720,* Manchester: Manchester University Press.

[75] Morris, P.W.G. 2013: 15. *Reconstructing Project Management,* Oxford: Wiley-Blackwell.

[76] Davis, H. 2006: 157. 'in the eighteenth century the roles of "architects", "builders", and "surveyors" were much more loosely defined than today', *The Culture of Building*, Oxford: Oxford University Press.

[77] Satoh, A. 1995: 37. *Building in Britain: The Origins of a Modern Industry*, London: Scholar Press.

[78] Cooney, E.W. 1955. The Origins of the Victorian Master Builders, *Economic History Review*, 2nd Series, No. 8, 167-76.

[79] Clark, L. 1992. *Building Capitalism – Historical Change and the Labour Process in the Production of the Built Environment*, London: Routledge.

[80] Murray, M. and Langford, D. 2003. *Construction Reports 1944-98,* Oxford: Wiley-Blackwell.

[81] Kingsford, P.W. 1973. *Builders and Building Workers*, London: Edward Arnold.

[82] Cooney, E.W. 1955. The Origins of the Victorian Master Builders, *Economic History Review*, 2nd Series, No. 8, 167-76.

[83] Satoh, A. 1995. *Building in Britain: The Origins of a Modern Industry*, London: Scholar Press.

[84] Powell, C.G. 1980: 29. *An Economic History of the British Building Industry 1815-1979*, London: The Architectural Press.

[85] Meikle, J. 2009: 27. *Thinking Big: A History of Davis Langdon,* London: Black Dog Publishing.

[86] Powell, C.G. 1980: 31. *An Economic History of the British Building Industry 1815-1979*, London: The Architectural Press.

[87] Davis, H. 2006: 158. *The Culture of Building*, Oxford: Oxford University Press.

[88] Satoh, A. 1995. *Building in Britain: The Origins of a Modern Industry*, London: Scholar Press. There are six chapters on nineteenth century technical advances, covering: stone, wood, bricks, components, pumps and lifting machinery.

Powell, C.G. 1980: 79-94. *An Economic History of the British Building Industry 1815-1979*, London: The Architectural Press, also discusses materials.

[89] Satoh, A. 1995: 96. *Building in Britain: The Origins of a Modern Industry*, London: Scholar Press.

[90] Satoh, A. 1995: 292. *Building in Britain: The Origins of a Modern Industry*, London: Scholar Press.

[91] Davis, H. 2006: 103. *The Culture of Building*, Oxford: Oxford University Press.

[92] Davis, H. 2006: 259-276. *The Culture of Building*, Oxford: Oxford University Press.

[93] Powell, C.G. 1980: 108. *An Economic History of the British Building Industry 1815-1979*, London: The Architectural Press.

[94] Fitch, J.M. 1973: 123. *American Building: The Historical Forces that Shaped It*, New York: Shocken Books.

[95] Rabeneck, A. 2011: 120. The invention of the building industry in Britain, *ArtefaCTos*, 4:1, 93-121.

[96] Loulakis, M.C. 2003: 49. *Design-build for the public sector*. New York: Aspen.

[97] Eccles, R. 1981. Bureaucratic versus craft administration: the relationship of market structure to the construction firm. *Administrative Science Quarterly, 26*(3), 449-469.

[98] Bowley, M. 1966: 352. *The British building industry- Four Studies in Response and Resistance to Change,* Cambridge: Cambridge University Press.

[99] Turin, D. 2003. Building as a process, *Building Research & Information*, 31:2, 180-187.

[100] Bresnen, M. 1990. *Organizing Construction: Project Organization and Matrix Management*. London: Routledge.

Davis, H. 2006: 164. *The Culture of Building*, Oxford: Oxford University Press. 'Today ... developers, with varying levels of civic responsibility, are at the centre of the building culture'.

[101] de Valence, G. and Huon, N. 1999. Procurement Strategies, in Best and de Valence (eds.) Building in Value: Pre Design Issues, Arnold Publishers, London. pp. 37-61.

[102] Dubois, A. and Gadde, L-E. 2002. The construction industry as a loosely coupled system: implications for productivity and innovation, *Construction Management & Economics*, 20 (7), 621-31.

[103] There have been similar issues and cost overruns in the construction of other more recent houses of parliament, in Australia, Germany and Scotland.

[104] de Valence, G. 2010. Innovation, procurement and construction industry development, *Australasian Journal of Construction Economics and Building*, 10 (4) 50-59.

[105] de Valence, G. 1997. Construction Industry Reform Strategies in Australia: CIDA and the CPSC, First International Conference on Construction Industry Development, National University of Singapore, Vol. 1, pp. 124-31. For the UK Latham (1992) and Egan (1998) are further discussed in chapter seven.

[106] Morris, P.W.G. 2013: 176. *Reconstructing Project Management,*

Oxford: Wiley-Blackwell.

[107] Satoh, A. 1995: 297-99. *Building in Britain: The Origins of a Modern Industry*, London: Scholar Press.

The First Industrial Revolution and the Industry Life-cycle

[108] Landes, D. S. 1972: 545. *The Unbound Prometheus: Technological Change from 1750 to the Present*, Cambridge: Cambridge University Press.

[109] Bowley, M. 1960. *Innovation in Building Materials*, London: Gerald Duckworth.

[110] Huxtable, A. L. 2008. *On Architecture: Collected Reflections on a Century of Change*, New York: Walker Publishing Company.

[111] Pfammatter, U. 2008. *Building the Future: Building Technology and Cultural History from the Industrial Revolution until Today*. Munich: Prestel Verlag.

[112] Peters, T. F. 1996: 178-202. *Building the Nineteenth Century*, Cambridge, Mass.: MIT Press.

[113] Peters, *Building the Nineteenth Century,* 1996: 244.

[114] Sichel, D. 2022. The Price of Nails Since 1695: A Window into Economic Change, *Journal of Economic Perspectives, Volume 36, Number 1: 125–150.* Provides price indexes for nails going back to 1695. The price of nails fell significantly relative to prices of an overall basket of consumption goods, with the real price of nails falling by a factor of about 10 from the late 1700s to the middle of the 20th century.

[115] The falling price of nails from hand-forged to cut to wire nails tracks broader changes across the US manufacturing sector and the evolution from artisanal to factory to continuous-process production technologies as described by Goldin, C, and Katz, L. 1998. The Origins of Technology-Skill Complementarity. *Quarterly Journal of Economics* 113 (3): 693–732.

[116] Fitch, J.M. 1973: 121. *American Building: The Historical Forces that Shaped It*, New York: Shocken Books.

[117] Jackson, K. 1987. *Crabgrass Frontier: The Suburbanization of the United States.* Oxford: Oxford University Press. Balloon-frame construction was much lower-cost than the traditional post-and-beam approach.

[118] Information on companies referred to has been sourced from their internet pages.

[119] Large, capital intensive, centrally controlled networks of energy, communication, transport and utilities are typically mature systems,

many developed during the second half of the nineteenth century. These are first-order systems, with second-order systems attached in a 'process of networking parts of different first-order systems for specific, macro-level social domains.' Braun, I. and Joerges, B. 1994: 26. How to recombine large technical systems: the case of European organ transplantation, in Summerton, J. (ed.) *Changing Large Technical Systems*, Boulder (CO): Westview.

Social studies of technological systems describe them as complex, heterogeneous, decentralized, self-organized, ungovernable, postmodern and transmodern systems, Coutard, O. 1999. (ed.) *The Governance of Large Technical Systems*, London: Routledge.

120 Hughes, T. P. 1987: 51 The evolution of large technological systems, in *The Social Construction of Technological Systems: New Directions in the Sociology and History of Technology*, W. E. Bjiker, T. P. Hughes, and T. J. Pinch (eds.), Cambridge, Mass.: MIT Press. 'During invention and development inventor-entrepreneurs solve critical problems; during innovation, competition, and growth manager-entrepreneurs make crucial decisions; and during consolidation and rationalization financier-entrepreneurs ... solve the critical problems associated with growth and momentum'.

121 Pfammatter, U. 2008. *Building the Future: Building Technology and Cultural History from the Industrial Revolution until Today*. Munich: Prestel Verlag.

122 Cody, J. 2003. *Exporting American Architecture 1870-2000*, London: Routledge.

Huxtable, A. L. 2008. *On Architecture: Collected Reflections on a Century of Change*, New York: Walker Publishing Company.

123 For a review of current research in concrete and mineral admixtures, glass and plastic, biological materials, nanotechnology, wood and other construction materials see Bamigboye, G. O., Davies, I., Nwanko, C., Michaels, T., Adeyemi, G. and Ozuor, O. 2019. Innovation in construction materials - A review, *IOP Conf. Ser.: Mater. Sci. Eng.* 640 012070

124 Syverson, C. 2019. Macroeconomics and Market Power: Context, Implications, and Open Questions, *Journal of Economic Perspectives*, 33, 3, 23–43.

Syverson, C. 2008. Markets: Ready-Mixed Concrete, *Journal of Economic Perspectives*, 22, 1, 217–233.

[125] Hughes saw large, modern technological systems evolving in stages: 'As systems mature, they acquire style and momentum.' The evolution of large technological systems, 1987: 50. A mature system is one where innovation and technological progress is managed as one part of the firm's focus on growth and competition.

[126] Coutard, O. 1999. The evolving forms of governance of large technical systems, in *The Governance of Large Technical Systems*, London: Routledge.

[127] Davis, H. 2006: 16. *The Culture of Building*, Oxford: Oxford University Press.

[128] Arthur, W. B. 1988. Competing Technologies, Increasing Returns and 'Lock-In' by Small Historical Events, *Economic Journal*, 99, 394, 116-131.

David, P. A. 1985. Clio and the Economics of QWERTY, *American Economic Review*, 75(2) 332-337.

[129] Landes, D.S. 1972: 318. *The Unbound Prometheus: Technological Change from 1750 to the Present*, Cambridge: Cambridge University Press. 'a serious bone of contention was the right of management to shift men about as needed.'

[130] Hughes, T. P. 1989: 495. *American Genesis: A Century of Invention and Technological Enthusiasm 1870-1970*, Chicago: University of Chicago Press.

[131] Gershenfeld, N. 2017: 108. The Science, and The Roadmap, in Gershenfeld, N., Gershenfeld, A. and Cutcher-Gershenfeld, J. *Designing Reality: How to Survive and Thrive in the Third Digital Revolution*, in New York: Basic Books. 95-116, and 159-182.

[132] Dubois and Gadde (2002) suggest a decentralised, project based structure can be a limiting factor on a firm's innovative potential, an argument also made by Gann and Salter (2000), Gann (2001, 2003) and Winch (1998).

[133] Bloom, N. and Van Reenen, J. 2010. Why Do Management Practices Differ across Firms and Countries? *Journal of Economic Perspectives*, 24(1), 203–224.

[134] Dosi, G., Freeman, C. and Fabiani, S. 1994. The Process of Economic Development: Introducing Some Stylized Facts and Theories on Technologies, Firms and Institutions, *Industrial and Corporate Change*, 3(1), 1-45.

[135] David, P. A. 1990. The Dynamo and the Computer: An Historical Perspective on the Modern Productivity Paradox, *American Economic Review*. 80, 355 - 361.

[136] Manuelli, R.E. and Seshadri, A. 2014. Frictionless Technology Diffusion: The Case of Tractors, *American Economic Review. 104 (4): 1368-91.*

[137] Allen, R. 2009. *The British Industrial Revolution in Global Perspective,* Cambridge: Cambridge University Press.

[138] Bresnen and Marshall, Understanding the diffusion and application of new management ideas in Bresnen, M. and Marshall, N. 2001. Understanding the diffusion and application of new management ideas in construction, *Engineering Construction and Architectural Management*, 8 (6), 335-45.

[139] Bonvillian, W. B. and Weiss, C. 2015: 177. *Technological Innovation in Legacy Sectors*, Oxford University Press USA.

[140] See annual reports for James Hardie Pty Ltd.

Industrialization and Modern Construction

[141] Cited by Wilkinson, T. 2014: 109. *Bricks and Mortals: Ten great buildings and the people they made*, New York: Bloomsbury Press.

[142] Davies, C. 2005. *The Prefabricated Home*, Reaktion Books.

[143] Herbert, G. 1984. *The Dream of the Factory-made House*, Harvard: MIT Press. The two famous architects behind the venture were Walter Gropius and Konrad Waschmann.

[144] https://lib.guides.umd.edu/kithouses

[145] Glaeser, E. 2011: 174-79. *Triumph of the City*, London: Macmillan.

[146] Dacy, D.C. 1965. Productivity and price trends in construction since 1947. *Review of Economics and Statistics*, 47(4), 406-411.

[147] Cassimatis, P.J. 1969. *Economics of the Construction Industry*, National Industrial Conference Board, New York.

Ive, G., Gruneberg, S., Meikle, J. and Crosthwaite, D. 2004. *Measuring the competitiveness of the UK construction industry* (Vol. 1). London: Department of Trade and Industry.

[148] Davis, H. 2006. *The Culture of Building*, Oxford: Oxford University Press.

[149] Stinchcombe, A.L. 1959. *Bureaucratic* and *craft administration* of production: A comparative study, *Administrative Science Quarterly*, 4, 168-87

[150] For a review of construction productivity research see de Valence, G. and Abbott, M. 2015. A review of the theory and measurement techniques of productivity in the construction industry, in Best, R. and Meikle, J. (eds.) *Measuring Construction: Prices, Output and Productivity*, London: Taylor & Francis, pp. 205-23.

[151] Rosenfielde, S. and Mills, D. 1979. Is construction technologically stagnant? In J.E. Lange & D.Q. Mills, (eds.) *The Construction Industry*, D.C. Heath and Company, Lexington MA.

[152] https://www.mlit.go.jp/sogoseisaku/1_6_hf_000154.html

[153] https://publications-cnrc.canada.ca/eng/view/ft/?id=52dc96d5-4ba0-40e6-98d2-8d388cba30cd

[154] https://www.legco.gov.hk/yr00-01/english/panels/plw/papers/plw0611-487e-scan.pdf

[155] Egan, J. 1998. *Rethinking construction,* London: Department of the Environment, Transport and the Regions.

[156] Egan, J. 1998. *Rethinking construction,* London: Department of the Environment, Transport and the Regions.

[157] *Modernising Construction*, National Audit Office, 2001, and *Accelerating Change,* Strategic Forum for Construction, 2002.

[158] Farmer, M. 2016. *Modernise or Die*, London: Construction Leadership Council.

[159] BRE, 2001. *Current practice and potential uses of prefabrication,* Building Research Establishment, Watford, U.K.

Taylor. M. 2010. A definition and valuation of the UK offsite construction sector, *Construction Management & Economics,* 28:8, 885-896. The paper is on definition and valuation of offsite construction in the UK using combined company financial accounts and records available from FAME with offsite construction supplier and manufacturer databases containing manufacturers and merchants as well as contractors.

[160] Gibb, A.G. 1999. *Off-site fabrication - pre-assembly, prefabrication and modularisation*, Whittles Publishing Services.

Pasquire, C.L. and Gibb, A.G.F. 2002. Considerations for assessing the benefits of standardisation and pre-assembly in construction, *Journal of Financial Management of Property and Construction*, 7(3), 151–61.

[161] Gann, D. 1996. Construction as a Manufacturing Process – similarities and differences between Industrialized housing and car production in Japan, *Construction Management & Economics* (14) pp 437 – 450.

[162] Farmer, M. 2016. *Modernise or die*, London: Construction Leadership Council.

[163] PATH, 2004. *Technology Roadmap: Advanced panelised construction, 2003 Progress Report.* Partnership for Advanced Technology in Housing (PATH), Department of Housing and Urban Development, Office of Policy Development and Research, Washington, D.C.

[164] https://www.ianvisits.co.uk/blog/2017/09/28/how-crossrail-is-using-3d-printing-in-its-stations/

[165] Thornton, R. 2002. *The Houses That Sears Built: Everything You Ever Wanted To Know About Sears Catalogue Homes,* Alton, Illinois Gentle Beam Publications.

[166] Polat, G., Arditi, D., Ballard, G., Mungen, U. 2006. Economies of on-site versus off-site fabrication of Rebar, *Construction Management & Economics* (24) 1185-1198.

[167] Cousineau, L. and Miura, N. 1998. *Construction Robots: The Search for New Building Technology in Japan.* ASCE Press.

[168] https://www.legalandgeneral.com/modular-homes/

[169] https://www.constructionenquirer.com/2020/11/30/lg-modular-homes-losses-exceed-100m/

[170] https://www.globalconstructionreview.com/news/modular-construction-unicorn-katerra-files-chapter/

[171] https://www.architectmagazine.com/technology/katerras-2-billion-legacy_o

[172] The OECD defined a platform as a 'digital service that facilitates interactions' in *An Introduction to Online Platforms and Their Role in the Digital Transformation,* 2019.

[173] Lim, M. K., Xiong, W. and Wang, C. 2021. Cloud manufacturing architecture: a critical analysis of its development, characteristics and future agenda to support its adoption, *Industrial Management & Data Systems.*

[174] Koc, E., Pantazis, E., Soibelman, L. and Gerber, D. 2020: 474. Emerging trends and research directions, in Sawhney, A., Riley, M. and Irizarry, J. (eds.) *Construction 4.0: An Innovation Platform for the Built Environment,* Abingdon: Routledge. 460 -75.

[175] Hall, D. M., Whyte, J. K. and Lessing, J. 2019: 15. Mirror-breaking strategies to enable digital manufacturing in Silicon Valley construction firms: a comparative case study, *Construction Management & Economics,*

[176] Varian, H. 2014. Beyond Big Data, *Business Economics,* 49(1): 27-31.

[177] Agrawal, Gans, and Goldfarb, *Prediction Machines: The Simple Economics of Artificial Intelligence*, 2018.

[178] Katz, M. L., and Shapiro, C. 1985. Network Externalities, Competition, and Compatibility, *American Economic Review*, 75(3): 424-440. Katz, M. L., and Shapiro, C. 1986. Technology Adoption in the Presence of Network Externalities, *Journal of Political Economy*, 94(4): 822-841. Katz, M. L., and Shapiro, C. 1992. Product Introduction with Network Externalities, *Journal of Industrial Economics*, 40(1): 55-83.

[179] Wei, W., Zhou, F. and Liang, P.F. 2020. Product platform architecture for cloud manufacturing. *Advances in Manufacturing* 8, 331–343.

[180] Lim, M. K., Xiong, W. and Wang, C. 2021. Cloud manufacturing architecture: a critical analysis of its development, characteristics and future agenda to support its adoption, *Industrial Management & Data Systems.*

[181] Winch, G. 2006. Towards a theory of construction as production by projects, *Building Research & Information,* 34:2, 154-163.

[182] https://www.census.gov/data/tables/time-series/econ/mhs/shipments.html

Construction 4.0, AI and Digital Fabrication

[183] Clarke, in a footnote to his 1973 revision of *Profiles of the Future.*

[184] Bloom, N., Van Reenen, J. and Williams, H. 2019. A toolkit of policies to promote innovation. *Journal of Economic Perspectives*, *33*(3), 163-84.

[185] https://www.weforum.org/centre-for-the-fourth-industrial-revolution/areas-of-focus

[186] https://constructionphysics.substack.com/p/a-brief-history-of-construction-startups?s=r

[187] https://constructionphysics.substack.com/p/a-brief-history-of-construction-startups?s=r

[188] Chaoran, C. 2020. Technology adoption, capital deepening, and international productivity differences, *Journal of Development Economics*, Volume 143, 102388.

[189] Australian 2021 procurement IPO Felix had local start-ups Buildxact, SiteMate, Mastt, Portt and VenderPanel with competing platforms. In the UK launched a procurement platform in 2021, mainly for the public sector, using framework agreements for building work valued between £250k to £10m.

[190] Sawhney, A., Riley, M. and Irizarry, J. (eds.) 2020: 5. *Construction 4.0: An Innovation Platform for the Built Environment*, Abingdon: Routledge. The chapters on each of these technologies give examples of current projects and research, with an evaluation on progress. The number of examples and their diversity in *Construction 4.0* highlight the widening scope of experimentation that is taking place, but also highlight their small scale and early stage of development. In the concluding chapter, despite the authors' optimism, the diffusion path they predict is 2050 and beyond.

[191] Kelly, 2016: 215. *The Inevitable: Understanding the 12 Technological Forces That Will Shape Our Future,* New York: Random House.

[192] Lavikka, R., Kallio, J., Casey, T. and Airaksinen, M. 2018 Digital disruption of the AEC industry: technology-oriented scenarios for possible future development paths, *Construction Management and Economics*, 36:11, 635-650.

[193] For details of machine learning methods see Mitchell, M. 2019. *Artificial Intelligence: A Guide for Thinking Humans,* New York: Farrar, Straus, and Giroux.

[194] The Obama administration released two reports on artificial intelligence at the end of 2016, with an overview of the state of play and background to the growing surge of interest and investment in AI: *Artificial Intelligence, Automation, and the Economy*, (December 2016), and *Preparing for the Future of AI*, (October 2016). At https://obamawhitehouse.archives.gov/blog/2016/12/20/artificial-intelligence-automation-and-economy.

[195] Mitchell, M. 2019. *Artificial Intelligence: A Guide for Thinking Humans,* New York: Farrar, Straus, and Giroux.

[196] Bianco, J. L., Fuchs, S., Parsons, M. and Riberinho, M. J. 2018. *Artificial Intelligence: Construction Technology's Next Frontier*, McKinsey & Co.

[197] The role of intelligent machines is to optimize outcomes by predicting potential points of failure. Agrawal, A., Gans, J. and Goldfarb, A. 2018. *Prediction Machines: The Simple Economics of Artificial Intelligence*, Harvard, Mass.: Harvard Business Review Press.

[198] Sanjayan, N. and Nematollahi, B. (eds.) 2019. *3D Concrete Printing Technology: Construction and building applications.* Butterworth-Heinemann.

[199] https://www.aniwaa.com

[200] Gershenfeld, N. 2012: 45. How to Make Almost Anything: The Digital Fabrication Revolution. *Foreign Affairs,* 91(6), pp. 43–57.

[201] Kaseman, K. and Graser, K. 2020: 187. Digital fabrication in the construction sector, in Sawhney, A., Riley, M. and Irizarry, J. (eds.) *Construction 4.0: An Innovation Platform for the Built Environment*, Abingdon: Routledge.

[202] https://fabfoundation.org/global-community/

[203] Lena-Acebo, F. J., and García-Ruiz, M. E. 2019. The FABLAB Movement: Democratization of Digital Manufacturing. In A. Guerra Guerra (Ed.), *Organizational Transformation and Managing Innovation in the Fourth Industrial Revolution.* IGI Global. pp. 125-142.

[204] Gershenfeld, N. 2017. The Science, and The Roadmap, in Gershenfeld, N., Gershenfeld, A. and Cutcher-Gershenfeld, J. *Designing Reality: How to Survive and Thrive in the Third Digital Revolution*, in New York: Basic Books. 95-116, also 159-182.

[205] Gershenfeld, N., Gershenfeld, A. and Cutcher-Gershenfeld, J. 2017: 7. *Designing Reality: How to Survive and Thrive in the Third Digital Revolution*, in New York: Basic Books.,

[206] Kaseman and Graser, Digital fabrication in the construction sector, 2020: 192.

[207] ARUP, *Designing with Digital Fabrication*, 2020: 8.

[208] Khoshnevis, B. 2004. Automated construction by contour crafting related robotics and information technologies, *Journal of Automation in Construction.* 13 (1) 5-19.

[209] Sanjayan, J. G., Nazari, A. and Nematollahi, B. 2019. *3D Concrete Printing Technology - Construction and Building Applications.* London: Elsevier.

[210] The trend towards integration of onsite and offsite production by contractors was first observed in the 1990s: 'The most important change is a redefinition of the construction firm to an organization that integrates the most important steps in the process of creating a new building. This means internalising the design and developing the appropriate technology and producing the building materials. Because of the integration, the new technology is not only in terms of erecting the building but in design and in developing new materials and in new ways of providing services. The products they sell will be solutions to accommodation problems, not the management services for erecting a predetermined building. The competition will be more akin to Schumpeter's creative destruction than to the cost-minimisation

models implied in the various management fads.' Runeson, G. 2000: 218. *Building Economics*, Deakin: Deakin University Press.

[211] Brynjolfsson, E. and McAfee, A. 2014: 186. *The Second Machine Age: Work, Progress and Prosperity in a Time of Brilliant Technologies*, New York: W. W. Norton & Co.

[212] Zolas, N. et al. 2020. *Advanced Technologies Adoption And Use By U.S. Firms: Evidence From The Annual Business Survey,* Nikolas, Zachary Kroff, Erik Brynjolfsson, Kristina McElheran, David N. Beede, Cathy Buffington, Nathan Goldschlag, Lucia Foster and Emin Dinlersoz. National Bureau of Economic Research, Cambridge, MA. Working Paper.

[213] International Standard 19650, https://www.iso.org/obp/ui/#iso:std:iso:19650:-1:ed-1:v1:en

[214] BCA, 2016. *BIM Essential Guide for DfMA* (Design for Manufacturing and Assembly), Building and Construction Authority, Singapore.

[215] Queensland Government 2018. Digital Enablement for Queensland Infrastructure - Principles for BIM Implementation, State of Queensland, Queensland.

[216] Jiang, R., Wu, C., Lei, X., Shemery, A., Hampson, K.D. and Wu, P. 2021. Government efforts and roadmaps for building information modelling implementation: lessons from Singapore, the UK and the US, *Engineering, Construction and Architectural Management.*

[217] https://www.ukbimalliance.org/wp-content/uploads/2021/06/UKBIMA-State-of-the-Nation-Survey-Report-2021.pdf

[218] UK Cabinet Office. *Government Construction Strategy 2016-2020.*

[219] https://www.ukbimframework.org/about/

[220] Lee, G. and Bormann, A. 2020. BIM policy and management, *Construction Management and Economics*, 38:5, 413-419. Links to the relevant documents for each country can be found in the article.

[221] Aiginger, K. and Rodrik, D. 2020. Rebirth of Industrial Policy and an Agenda for the Twenty-first Century, *Journal of Industry, Competition and Trade*, 20:189–207.

The Built Environment and Industry Policies

[222] Powell, C. G. 1980: 131. *An Economic History of the British Building Industry 1815-1979*, London: The Architectural Press. Describing the UK Government's role as both a client and policy maker in the

1950s and 1960s.

[223] Acemoglu, D. 2002. Directed Technical Change. *Review of Economic Studies*. 69: 781-809.

[224] Aghion, P., A. Dechezleprêtre, D. Hemous, R. Martin, and J. Van Reenen. 2016. Carbon Taxes, Path Dependency and Directed Technical Change: Evidence from the Auto Industry. *Journal of Political Economy*. 124: 1-51.

[225] Porter, M. E. 1985. *Competitive Advantage*, New York: Free Press.

[226] Bloom, N., Van Reenen, J. and Williams, H. 2019. A toolkit of policies to promote innovation. *Journal of Economic Perspectives*, *33*(3), 163-84.

[227] Warwick, K. 2013. Beyond Industrial Policy: Emerging Issues and New Trends, *OECD Science, Technology and Industry Policy Papers*, No. 2, OECD Publishing, Paris. Gives 17 different definitions of industrial policy with their sources on page 15.

[228] Chang, H-J. and Andreoni, A. 2020. Industrial Policy in the 21st Century, *Development and Change,* 51(2): 324–351.

[229] Acemoglu, D., P. Aghion, L. Bursztyn, and D. Hemous. 2012. The Environment and Directed Technical Change. *American Economic Review*. 102(1): 131-166.

[230] Gruneberg, S. 2020. *A Strategic Approach to the UK Construction Industry,* London: Taylor and Francis.

[231] Murray, M. and Langford, D. 2003: 7. *Construction Reports 1944-98,* Oxford: Wiley-Blackwell.

[232] Green, S.D. 2011. *Making Sense of Construction Improvement*, Wiley-Blackwell, Oxford.

Green, S.D. and May, S.C. 2003. Re-engineering construction: going against the grain, *Building Research & Information,* 31(2), 97-106

[233] Fernie, S., Leiringer, R. and Thorpe, T. 2006. Rethinking change in construction: a critical perspective. *Building Research & Information,* **34**(2), 91-103.

[234] Wolstenholme, A. 2009: 8. *Never Waste a Good Crisis: A review of progress since rethinking construction and thoughts for our future*, London: Construction Excellence.

[235] Davis, H. 2006. *The Culture of Building*, Oxford: Oxford University Press.

[236] Simmons, R. 2015. Constraints on evidence-based policy: insights from government practices, *Building Research & Information*, 43:4, 407-419.

[237] Carroll, P. 2010. Does regulatory impact assessment lead to better policy? *Policy & Society*, 29:2, 113-122.

[238] Wond, T. and Macaulay, M. 2010. Evaluating local implementation: An evidence-based approach. *Policy & Society*, 29:2, 161-169.

[239] See the discussion in the Appendix.

[240] Greenwood, D. 2019. The adoption of building information modelling as a measure of international construction innovation' in Gruneberg, S. (ed.) 2019. *Global Construction Data*, London: Taylor & Francis. 111 -125.

[241] http://bim-level2.org/globalassets/pdfs/bim-level-2-introduction-to-commercial-documents.pdf

[242] Links to the UK Government's Construction 2011 and 2025 policy documents and the Government Construction Strategy 2016-2020: https://www.gov.uk/government/uploads/system/uploads/attachment_data/file/61152/Government-ConstructionStrategy_0.pdf

https://www.gov.uk/government/uploads/system/uploads/attachment_data/file/210099/bis-13-955-construction-2025-industrial-strategy.pdf

https://www.gov.uk/government/uploads/system/uploads/attachment_data/file/510354/Government_Construction_Strategy_2016-20.pdf

[243] https://www.ukbimframework.org/about/

[244] https://constructionmanagermagazine.com/covid-aids-bim-adoption-but-barriers-remain/

[245] https://www.ukbimalliance.org/wp-content/uploads/2021/06/UKBIMA-State-of-the-Nation-Survey-Report-2021.pdf

[246] https://constructionglobal.com/digital-construction/which-countries-are-leading-adoption-bim

[247] Although there is an economic activity called construction in the SIC the characteristics of the three divisions makes them different industries. The manufacturing SIC includes glass, wood products, steel, plastics and concrete, but they are regarded as separate industries and are not grouped together under a construction products SIC. An industry policy for the steel industry is not thought to apply to plastics or concrete because it is not relevant to those industries. The same applies to the differences between residential building, non-residential building and engineering construction.

[248] Industry culture is a complex outcome of social (Beamish and Biggart 2012), institutional (Davis 1999) and economic (Powell 1990) factors. Because of the range and dynamic interplay of those factors it is not an appropriate target for industry policy, as the history of construction reform efforts that argued cultural change was necessary for industry improvement in the UK shows.

[249] For example Boston Consulting Group, *Digital in Engineering and Construction: The Transformative Power of Building Information Modeling*, 2017.

[250] Pfammatter, U. 2008. *Building the Future: Building Technology and Cultural History from the Industrial Revolution until Today*. Munich: Prestel Verlag.

Davis, H. 2006. *The Culture of Building*, Oxford: Oxford University Press.

[251] For details and a history of the ISO, and for several standards, see Rich, N. and Malik, N. 2019. *International Standards for Design and Manufacturing: Quality Management and International Best Practice*, London: Kogan Page.

[252] https://codes.iccsafe.org

[253] Miao and Popp studied innovative responses to three natural disasters: earthquakes, flooding, and drought. Based on the frequency and location of natural disasters and a panel of patent data from 1974-2009, they find that a billion dollars of damage in a country from natural disasters increased innovation by 18 to 39 percent. Miao, Q. and D. Popp. 2014. Necessity as the Mother of Invention: Innovative Responses to Natural Disasters. *Journal of Environmental Economics and Management*. 68(2): 280- 295.

[254] The effectiveness of standards encouraging the use and diffusion of environmental technologies like renewable energy and energy efficiency are reviewed by Vollebergh, H.R.J. and E. van der Werf. 2014. The Role of Standards in Eco-Innovation: Lessons for Policymakers. *Review of Environmental Economics and Policy*. 8: 230–248.

[255] On the importance of sustainability and the contribution construction economics can make to decarbonization see Myers, D. 2017. *Construction Economics: A new approach*, 4th Ed. London: Spon Press.

[256] For a comprehensive review see Popp, D. 2019. Environmental Policy and Innovation: A Decade of Research. CESifo Working Paper No.

7544 (on SSRN). This updates the earlier review: Popp, D., R. Newell and A.B. Jaffe 2010. Energy, the Environment, and Technological Change. In *Handbook of the Economics of Innovation: vol. 2*, Hall, B. and Rosenberg, N. (eds.), Academic Press/Elsevier, 873-937.

257 Armstrong, A., Wright, C., Ashe, B., and Nielsen, H. 2017. Enabling Innovation in Building Sustainability: Australia's National Construction Code, *Procedia Engineering*, Volume 180, 320-330.

258 Leibwicz, B. D. 2017. Effects of urban land-use regulations on greenhouse gas emissions, *Cities*, 70, 135-152.

259 United Nations Environment Programme. 2021: 59. *Global Status Report for Buildings and Construction: Towards a Zero-emission, Efficient and Resilient Buildings and Construction Sector*. Nairobi https://globalabc.org/sites/default/files/2021-10/GABC_Buildings-GSR-2021_BOOK.pdf

260 https://www.citiesalliance.org/draft-international-standard-determining-greenhouse-gas-emissions-cities

261 In Europe national building codes set energy requirements to induce innovation. One study found positive effects from policies to improve energy efficiency in new residential buildings, such as energy-efficient boilers and improved insulation, lighting and materials. Prices were found to have an effect on innovation for visible technologies such as boilers and lighting, but not for less-visible technologies such as insulation that are installed by builder. Noailly, J. 2012. Improving the Energy Efficiency of Buildings: The Impact of Environmental Policy on Technological Innovation. *Energy Economics*. 34: 795-806.

262 https://www.imt.org/wp-content/uploads/2022/02/SPIPA_EU_rev6_508.pdf. p. 11.

263 https://www.imt.org/resources/mapping-us-policy-on-energy-efficiency-in-buildings/

264 https://www.imt.org/resources/mapping-us-policy-on-energy-efficiency-in-buildings/

265 https://www.forbes.com/sites/energyinnovation/2020/12/02/a-powerful-yet-underused-climate-tool-building-codes/?sh=567e3d0ed978

266 https://www.energystar.gov/buildings/benchmark/understand_metrics/how_score_calculated

Three Pathways to Future Construction

[267] Groak, S. 1992: 169. *The Idea of Building*, London: Spon.

[268] Agrawal, A., Gans, J. and Goldfarb, A. 2018. *Prediction Machines: The Simple Economics of Artificial Intelligence*, Harvard, Mass.: Harvard Business Review Press.

[269] Sovacool, B. K., Lovell, K. and Ting, M. B. 2018. Reconfiguration, Contestation, and Decline: Conceptualizing Mature Large Technical Systems, *Science, Technology, & Human Values*, Vol. 43(6) 1066-1097.

Housing is used as an example in Bonvillian, W. B. and Weiss, C. 2015. *Technological Innovation in Legacy Sectors*, Oxford University Press USA.

[270] Greenstein, S. 2015. *How the Internet Became Commercial: Innovation, Privatization, and the Birth of a new Network,* Princeton University Press.

[271] Forman, C., Goldfarb, A. and Greenstein, S. 2018. How Geography Shapes – and is Shaped by – the Internet, in (ed.) Clark, G. Feldman, M. Gertler, M. and Wojcik, D. *The New Oxford Handbook of Economic Geography,* Oxford University Press. 269-285.

[272] Jin, W. and McElheran, K. 2018. Economies before Scale: Learning, Survival, and Performance of Young Plants in the Age of Cloud Computing. SSRN Electronic Journal: https://papers.ssrn.com/sol3/papers.cfm?abstract_id=3112901

McElheran, K. 2015. Do Market Leaders Lead in Business Process Innovation? The Case(s) of E-Business Adoption, *Management Science*, 61(6), 1197-1216.

[273] Destafano, T., Kneller, R., and Timmins, J. 2018. Broadband Infrastructure, ICT use and firm performance: Evidence for UK Firms, *Journal of Economic Behavior & Organization,* 155(C), pp 110-139.

[274] Weitzman, M. 1980. The "ratchet principle" and performance incentives. *Bell Journal of Economics* 11: 302–8.

Weitzman, M. 1976. The new Soviet incentive model. *Bell Journal of Economics,* 7: 251-57.

[275] Ickes, B.W. and Samuelson, L. 1987. Job transfers and incentives in complex organizations: Thwarting the ratchet effect. *RAND Journal of Economics* 18, no. 2 (Summer): 275–86.

[276] Park, W. and Chapin, W. 1992. *Construction Bidding: Strategic pricing*

for profit, 2nd ed. New York: Wiley.

[277] Kagel, J. H. and Levin, D. 1986. The Winner's Curse and Public Information in Common Value Auctions. *American Economic Review*, 894–920.

[278] Edgerton, D. 2007. *The Shock of the Old: Technology and Global History since 1900*. Oxford: Oxford University Press.

[279] Andrews, D., C. Criscuolo and P. N. Gal, 2015. *Frontier Firms, Technology Diffusion and Public Policy: Micro Evidence from OECD Countries,* OECD Productivity Working Papers, 2015-02, Paris: OECD.

Bloom, N. and Van Reenen, J. 2010. Why Do Management Practices Differ across Firms and Countries? *Journal of Economic Perspectives*, 24(1), 203–224.

[280] WEF, 2016. *Shaping the Future of Construction: A Breakthrough in Mindset and Technology*, World Economic Forum and the Boston Consulting Group, Geneva.

[281] Sutton, J. 1999. *Technology and Market Structure*, Cambridge Mass.: MIT Press.

[282] Runeson, G. and de Valence, G. 2008. The New Construction Industry, in Ruddock, L. (ed.) *Economics for the Modern Built Environment*, London: Taylor and Francis, 199-211.

[283] Acemoglu D. and Restrepo, P. 2019. Artificial Intelligence, Automation, and Work, in Agrawal, A. Gans, J. and Goldfarb, A. (eds.) *The Economics of Artificial Intelligence: An Agenda*, London: The University of Chicago Press. 197-236

[284] Bianco, J. L., Fuchs, S., Parsons, M. and Riberinho, M. J. 2018. *Artificial Intelligence: Construction Technology's Next Frontier*, McKinsey & Co.

[285] Kamien, M. L. and Schwartz, N. L. 1982. *Market Structure and Innovation*, Cambridge: Cambridge University Press.

[286] Dosi, G. 1982. Technological paradigms and technological trajectories. A suggested interpretation of the determinants and directions of technical change, *Research Policy*, 11(3):147-162

Conclusion

[287] Giuseppe di Lampedusa, *The Leopard*, 1958. London: Fontana. 1963: 29.

[288] A good example of this negative view of construction is McKinsey, 2017. *Reinventing construction through a productivity revolution*, McKinsey

Global Institute.

[289] Lipsey, R.G., Carlaw, K. I. and Bekar, C. T. 2005. *Economic Transformations: General Purpose Technologies and Long-term Economic Growth*, Oxford: Oxford University Press, found 30 years was needed for development of a GPT, and diffusion took another 20 years.

Mowery, D.C. and Rosenberg, N. 1998. *Paths of Innovation: Technological Change in 20th Century America*, Cambridge: Cambridge University Press.

[290] Bainbridge, L. 1983. Ironies of automation. *Automatica, 19* (6), 775-779.

[291] Munn, L. 2022. *Automation is a Myth*, Stanford University Press.

[292] Rodrik, D. and Stantcheva, S. 2021. Fixing capitalism's good jobs problem, *Oxford Review of Economic Policy*, 37: 824-837.

Appendix: The Built Environment Sector and Construction Statistics

[293] Ive and Gruneberg (2000: 9) argued 'A *narrow* definition of the construction industry includes only those firms undertaking on-site activity' whereas a *broad* definition includes many firms from other industries involved in production of the built environment.

[294] Cruz, S. C. and Teixeira, A. A. 2010. The evolution of the cluster literature: Shedding light on the regional studies–regional science debate. *Regional Studies*, 44(9), 1263–1288.

[295] Construction value added is 13 percent of global GDP. World Bank, 2013. *Measuring the Real Size of the World Economy: The framework, Methodology and Results of the International Comparison Program – ICP*, Washington DC: World Bank.

[296] de Valence, G. 2019. Reframing construction within the Built Environment Sector, *Engineering, Construction and Architectural Management*, 26 (5) 740-745. See Turin (1969), Strassman (1970) and Groak (1992) for the origins and development of this idea.

[297] Ofori, G. 1990: 23. *The Construction Industry: Aspects of its Economics and Management*, Singapore: Singapore University Press.

[298] ABS 2009, *Information Paper: Introduction of revised international statistical standards in the Australian Tourism Satellite Account*, Cat. no. 5249.0.55.002, Australian Bureau of Statistics.

[299] Using what they called a 'meso-economic' approach (i.e. between micro and macroeconomics) Carassus *et al.* (2006) compared the size of the

'construction sector system' in seven countries, again using a wide definition to include property management, repair and maintenance, and the institutional actors involved. Ruddock and Ruddock (2009) followed this approach and used SIC data on it to get estimates for the construction sector in 20 European countries, which ranged between 12 and 22 per cent of GDP, with an average of about 17 per cent. Squicciarini and Asikainen (2011) extended the scope of their 'wide industry' considerably further than Pearce (2003).

300 UN, 2008: 3. *International Standard Industrial Classification of all Economic Activities*. New York: United Nations Publications.

301 UN, 2008: 175-179. *International Standard Industrial Classification of all Economic Activities*. New York: United Nations.

302 Meikle, J. and Gruneberg, S. 2015: 126. Measuring and comparing construction internationally. In: Best, R, and Meikle, J. (eds.) *Measuring Construction: Prices, Output and Productivity*, Abingdon: Routledge.

303 There are four studies that have quantified the relationship between the narrow and wider definitions of construction, two for the UK, (Ive and Gruneberg 2000, Pearce 2003) and two for Australia by de Valence in 2001 (based on AEGIS 1999), and 2022, Output and employment in Australian built environment industries 2007-19, in Best, R. and Meikle, J. (eds.) *Describing Construction: Industries, projects and firms*, London: Taylor and Francis

304 de Valence, G. and Abbott, M. 2015. A review of the theory and measurement techniques of productivity in the construction industry, in Best, R. and Meikle, J. (eds.) *Measuring Construction: Prices, Output and Productivity*, London: Taylor & Francis, pp. 205-23.

305 'Arranged throughout this [production system] — from resources to suppliers and components to innovation, through production, to distribution, retail, and life cycle — a great array of skills and firms. Many of these would be counted as services in economic statistics, but are actually tied to production.' Bonvillian, W. B., and Weiss, C. 2015: 45. *Technological Innovation in Legacy Sectors*, New York: Oxford University Press.

306 Briscoe, G. 2006: 220. How useful and reliable are construction statistics? *Building Research & Information, 34*(3), 220-229. And Cannon, J. 1994. Lies and construction statistics. *Construction Management and Economics, 12*(4), 307-313.

[307] UN, 2008. *System of National Accounts 2008*, New York: United Nations. Chapter 29 is on satellite accounts. http://unstats.un.org/unsd/nationalaccount/docs/SNA2008.pdf.

[308] The Australian tourism industry has an annual tourism satellite account produced by the ABS (ABS 5249), estimating the contribution to GDP at around three percent.

[309] https://unece.org/DAM/stats/documents/ece/ces/2019/ECE_CES_2019_18-G1910795E.pdf

[310] https://www.oecd-ilibrary.org/social-issues-migration-health/a-system-of-health-accounts_9789264116016-en

[311] https://www.bea.gov/research/papers/2020/measuring-small-business-economy

[312] https://www.bea.gov/data/special-topics/space-economy

[313] https://www.oecd-ilibrary.org/economics/developing-thematic-satellite-accounts_b833cbfa-en

[314] In selecting a number of industries of special interest 'It is common practice to refer to such groupings of industries as "sectors" even though they do not constitute institutional sectors as the term is used in the SNA. The SNA does not try to provide specific and precise criteria for the definition of what identifies a key sector or activity….. in some important cases, such as tourism and environmental protection activities, the process of identification of characteristic and connected products is complex because not all the relevant activities and products appear in the central framework classifications.' OECD, 2000. *A System of Health Accounts*, OECD Publishing, Paris.

Characteristic products are those that are typical of the field, for construction characteristic products are buildings and structures, project management and other professional services. Connected goods and services includes expenditure on products that are not typical and are classified to other product categories. In construction quarrying, manufactured products and transportation of materials and components may be considered connected.

[315] Sveikauskas, L., Rowe, S., Mildenberger, J., Price, J. and Young, A. 2018. Measuring productivity growth in construction, *Monthly Labor Review,* U.S. Bureau of Labor Statistics.

Sveikauskas, L., Rowe, S., Mildenberger, J., Price, J. and Young, A. 2016. Productivity growth in construction, *Journal of Construction*

Engineering and Management, vol. 142, no. 10.

Sveikauskas, L., Rowe, S., Mildenberger, J., Price, J. and Young, A. 2014. Productivity growth in construction, BLS Working Paper 478. https://www.bls.gov/osmr/research-papers/2014/ec140090.htm. The working paper has much more detail on how the estimates were prepared.

[316] In a footnote Sveikauskas et al. 2014: 26 'examine the sensitivity of these assumptions and also calculated the overall growth of labor input using weights of 1.0 for new construction, 1.5 for additions and alterations, and 2.0 for maintenance and repair. By giving less weight to additions or maintenance, these calculations assign more subcontractor labor to new construction. However, the alternative assumption increases overall total labor input growth by only 0.1 percent per year or less.'

[317] Sveikauskas, L., Rowe, S., Mildenberger, J., Price, J. and Young, A. 2014: 25-6. Productivity growth in construction, Bureau of Labor Statistics Working Paper 478.

[318] For a review of construction productivity research see de Valence, G. and Abbott, M. 2015. A review of the theory and measurement techniques of productivity in the construction industry, in Best, R. and Meikle, J. (eds.) *Measuring Construction: Prices, Output and Productivity*, London: Taylor & Francis, pp. 205-23.

[319] Rosenfielde, S. and Mills, D. 1979. Is construction technologically stagnant? In J.E. Lange & D.Q. Mills, (eds.) *The Construction Industry*, D.C. Heath and Company, Lexington MA.

[320] Issues with construction statistics were recognised as early as 1969 by Cassimatis, P. *Economics of the Construction Industry*, National Industrial Conference Board, New York. Argued again for low levels of US productivity growth by Scriver and Bowlby 1985.

[321] On UK construction statistics see Green, B. 2019. The challenges of measuring British construction output. In R. Best and J. Meikle (Eds.), *Accounting for Construction: Frameworks, Productivity, Cost and Performance,* New York: Routledge. pp 46-73.

[322] Green, S.D., Fernie, S., Weller, S. 2005. Making sense of supply chain management: a comparative study of aerospace and construction, *Construction Management & Economics*, 23(6), 579-593.

[323] Latham, M. 1994. *Constructing the Team*. Final Report of the Government/ Industry Review of Procurement and Contractual Arrangements in the UK Construction Industry, London: HMSO.

[324] Gann, D. 1996. Construction as a Manufacturing Process – similarities and differences between Industrialized housing and car production in Japan, *Construction Management & Economics* (14) 437 – 450.

[325] Despite the importance of repair and maintenance, only Canada has an annual business capital and repair expenditures survey. Between 2006 and 2016 construction R&M by firms averaged nine percent of their total capital expenditure, or around 1.2 percent of GDP, ranging between one percent of GDP in 2006 and 1.3 percent in 2012. Statistics Canada. Table: 34-10-0035-01 *Capital and repair expenditures, non-residential tangible assets.*

[326] UK construction turnover was 50% of all manufacturing turnover and almost 60% of all manufacturing employment in 2018, and much larger than any of the individual manufacturing industries. The largest manufacturing industry was motor vehicles, with 22% of construction turnover and 5.5% of construction employment. And it is manufacture of motor vehicles that has been compared with construction (Gann 1996, Egan 1998, Green et al. 2005).

[327] de Valence, G. 2022. On the measurement and characteristics of construction firms in theory and practice, in Best, R. and Meikle, J. (eds.) *Describing Construction: Industries, projects and* firms, London: Taylor and Francis.

[328] https://www.bls.gov/lpc/construction.htm. For details on methods and issues see Sveikauskas, L., Rowe, S., Mildenberger, J., Price, J. and Young, A. 2018. Measuring productivity growth in construction, *Monthly Labor Review,* U.S. Bureau of Labor Statistics.

[329] Yu, M. and Ive, G. 2015. Review of construction cost and price indices in Britain, in Best, R. and Meikle, J. (eds.) *Measuring Construction: Prices, output and productivity*, Oxford: Taylor and Francis, pp. 79-112.

[330] https://www.bls.gov/lpc/construction.htm.

Made in United States
Troutdale, OR
11/03/2026

56844464R00153